elementals

i. earth

volume i

earth

Hannah Eisler Burnett & Kristi Leora Gansworth, editors
Nickole Brown & Craig Santos Perez, poetry editors

Gavin Van Horn & Bruce Jennings, series editors

**Humans
& Nature
press**

Humans and Nature Press, Libertyville 60030
© 2024 by Center for Humans and Nature

For more information, contact Humans & Nature Press, 17660 West Casey Road, Libertyville, Illinois 60048.
Printed in the United States of America.

Cover and slipcase design: Mere Montgomery of LimeRed, https://limered.io

ISBN-13: 979-8-9862896-3-2 (paper)
ISBN-13: 979-8-9862896-4-9 (paper)
ISBN-13: 979-8-9862896-5-6 (paper)
ISBN-13: 979-8-9862896-6-3 (paper)
ISBN-13: 979-8-9862896-7-0 (paper)
ISBN-13: 979-8-9862896-2-5 (set/paper)

Names: Burnett, Hannah Eisler, editor | Gansworth, Kristi Leora, editor | Brown, Nickole, poetry editor | Perez, Craig Santos, poetry editor | Van Horn, Gavin, series editor | Jennings, Bruce, series editor

Title: Elementals: Earth, vol. 1 / edited by Hannah Eisler Burnett and Kristi Leora Gansworth

Description: First edition. | Libertyville, IL: Humans and Nature Press, 2024 | Identifiers: LCCN 2024902616 | ISBN 9798986289632 (paper)

Copyright and permission acknowledgments appear on page 153.

Humans and Nature Press
17660 West Casey Road, Libertyville, Illinois 60048

www.humansandnature.org

Printed by Graphic Arts Studio, Inc. on Rolland Opaque paper. This paper contains 30% post-consumer fiber, is manufactured using renewable energy biogas, and is elemental chlorine free. It is Forest Stewardship Council® and Rainforest Alliance certified.

contents

Gathering: Introducing
the Elementals *Series*

Gavin Van Horn and Bruce Jennings

T hunderous, cymbal-clashing waves. Dervish winds whip-
ping across mountain saddles. Conflagrations of flame lick-
ing at a smoke-filled sky. The majesties of desert sands and
wheat fields extending beyond the horizon. What riotous conflu-
ence of sound, sight, smell, taste, and touch breaches your imag-
ination when you call to mind the elementals? Yet the elementals
may enter your thoughts as subtler, quieter presences. The gentle
burbling of clear creek water. The rich loamy soil underfoot on a
trail not often followed. A pine-scented breeze wafting through a
forest. The inviting warmth of a fire in the hearth.

This last image of the hearth fire is apropos for the five vol-
umes that constitute *Elementals*. The fire, with its gift of collective
warmth, is a place to gather and cook together, and not least of all
a place that invites storytelling. And in stories the elementals can
be imagined as a better way of living still to be attained.

The essays and poems in these volumes offer a wide variety of
elemental experiences and encounters, taking kaleidoscopic turns
into the many facets of earth, air, water, and fire. But this series
ventures beyond good storytelling. Each of the contributions in the
pages you now hold in your hands also seeks to respond to a ques-
tion: What can the vital forces of earth, air, water, and fire teach
us about being human in a more-than-human world? Perhaps
this sort of question is also part of experiencing a good fire, the
kind in which we can stare into the sparks and contemplate our

lives, releasing our imaginations to possibilities, yet to be fulfilled but still within reach. The elementals live. Thinking and acting through them—in accommodation with them—is not outmoded in our time. On the contrary, the rebirth of elemental living is one of our most vital needs.

For millennia, conceptual schemes have been devised to identify and understand aspects of reality that are most essential. Of enduring fascination are the four material elements: earth, air, water, and fire. For much longer than humans have existed—indeed, for billions of years—the planet has been shaped by these powerful forces of change and regeneration. Intimately part of the geophysical fluid dynamics of the Earth, all living systems and living beings owe their existence and well-being to these elemental movements of matter and flows of energy. In an era of anthropogenic influence and climate destabilization, however, we are currently bearing witness to the dramatic and destructive potential of these forces as it manifests in soil loss, rising sea levels, devastating floods, and unprecedented fires. The planet absorbs disruptions brought about by the activity of living systems, but only within certain limits and tolerances. Human beings collectively have reached and are beginning to exceed those limits. We might consider these events, increasing in frequency and intensity, as a form of pushback from the elementals, an indication that the scale and scope of human extractive behaviors far exceeds the thresholds within which we can expect to flourish.

The devastating unleashing of elemental forces serves as an invocation to attend more deeply to our shared kinship with other creatures and to what is life-giving and life-nurturing over long-term time horizons. In short, caring about the elementals may also mean caring for them, taking a more care-full approach to them in our everyday lives. And it may mean attending more closely to the indirect effects of technological power employed at the behest of rapacious desires. Unlike more abstract notions of nature or numerical data about species loss, air measurements in parts per million, and other indicators of fraying planetary relations, the elementals can ground our moral relations in something tangible and close at hand—near as our next breath, our next meal, our next drink, our next dark night dawning to day.

For each element, the contributors to this series—drawing from their diverse geographical, cultural, and stylistic perspectives—explore and illuminate practices and cosmovisions that foster reciprocity between people and place, human and nonhuman kin, and the living energies that make all life possible. The essays and poems in this series frequently approach the elements from unexpected angles—for example, asking us to consider the elemental qualities of bog songs, the personhood of rivers, yogic breath, plastic fibers, coal seams, darkness and bird migration, bioluminescence, green burial and mud, the commodification of oxygen, death and thermodynamics, and the healing sociality of a garden, to name only a few of the creatively surprising ways elementals can manifest.

Such diverse topics are united by compelling stories and ethical reflections about how people are working with, adapting to, and cocreating relational depth and ecological diversity by respectfully attending to the forces of earth, air, water, and fire. As was the case in the first anthology published by the Center for Humans and Nature, *Kinship: Belonging in a World of Relations*, the fifth and final volume of *Elementals* looks to how we can live in right relation, how we can *practice* an elemental life. There you'll find the elements converging in provocative ways, and sometimes challenging traditional ideas about what the elements are or can be in our lives. In each of the volumes of *Elementals*, however, our contributors are not simply describing the elementals; they are also always engaging the question, How are we to live?

In a sense, as a collective chorus of voices, *Elementals* is a gathering; we've been called around the fire to tell stories about what it means to be human in a more-than-human world. As we stare into this firelight, recalling and hearing the echoing voices of our living planet, we stretch our natural and moral imaginations. Having done so, we have an opportunity to think and experiment afresh with how to live with the elementals as good relatives. The elementals set the thresholds; they give feedback. Wisdom—if defined as thoughtful, careful practice—entails conforming to what the elements are "saying" and then learning (over a lifetime) how to better listen and respond. Pull up a chair, or sit on the ground near the crackling glow; we'll gaze into the fire together and listen—to the stories that shed light and comfort, to the stories that discombobulate and help us see old things in a new way, to the stories that bring us back to what matters for carrying on together.

Introduction: Living with Our Many Earths

Hannah Eisler Burnett and Kristi Leora Gansworth

Welcome to Earth. We are happy to be sharing space and time with you here among the pages of *Earth*, as we are momentarily transported to wherever you are as you read these words. Perhaps the sun is shining through a window, and you are reminded of Earth's celestial connections. Perhaps it's cloudy, you can smell the decomposing leaves and logs of late autumn, and you are reminded of Earth's regenerative rhythms. Your surroundings—your context—will transform your experience of this text.

Like you, as coeditors, we each bring to this work our own histories, associations, and relationships with the Earth: Hannah grew up in Mississippi alongside the Gulf of Mexico, and her relationship with Earth has been animated by the memories and omissions of five generations of maternal, white, settler ancestors—all of whom called Mississippi home—and by the faraway recollections of her paternal Jewish family and their experiences of the places they fled. She is relatively new to where she is. Leora's roots are deeply embedded in the Earth of North America, along with the many inheritances that come from her Anishinabe and Onkwehonwe ancestors—human, spirit, plant, animal, bird, fish, and others.

When considering the idea of Earth-as-element, alongside the entities of fire, water, and air, the challenge of multiplicity immediately presents itself. Earth carries a range of relational meanings. Wildly multiscalar, Earth is at once planet, place, and material:

the word *Earth* invokes the light-years of distance between celestial bodies and the micron of space held by a wisp of spider silk. *Earth* connotes a relationship to the universe, to planet, to soil, and to season.

As a planet, the Earth forms humans' experience of time through its interactions with the sun and moon. The seasons, the tides, the days, and the nights are our experiences of Earth's movement. As living and undying geological matter, earth generates life from death and decomposition, bookending life histories and layering them into the sedimentary composition of mountains, gorges, and cliffs on land and under the sea, imbuing deep time into material space.

The Earth holds every other element together, providing a home for quotidian, lowercase *earth*—soil, stone, and dirt—and also to fire, to water, to air, and to the human meaning making that weaves all of those together. Earth's minerals are shaped by fire, water, air, and living beings. Her atmosphere feeds fire, her surface and subsurface cradle water and release it back to the sky under the influence of the sun's heat and light. These processes are inseparable from the emergence and survival of human life. Earth's minerals swirl in the bloodstreams and cavities of human and other bodies, an intricate balance of symbiosis and interdependence, a connector between realms.

Earth is often referred to in some cultures as the great Mother, a vast and cavernous generator for diverse forms of matter—water, medicine plants, poison plants, trees, fish, arthropods, birds, grasses, flowers, fruits, predators, prey, mammals, and countless other species, and the interactions among them. There is massive complexity flowing from Earth, which unites and nourishes. Earth is the host for changing times, conditions, and challenges while continuing to hold steady and morph, generating fertile conditions that enable human and biodiverse life to persist.

Many may describe Earth as "the environment," yet this volume and its contributors challenge the idea that the Earth is an entity outside of us. Earth created the conditions of being for all life, and we are composed of the elements that make up the Earth. Our survival and thriving is the survival and thriving of Earth. As toxic by-products of human industrial activity waft through the air, we breathe them in. As the same industries send waste into the waters of Earth, we drink and reincorporate these waters into ourselves. We are all implicated in the Earth's well-being. There is no separation between our bodies and the body of Earth. Humanity is not a figure that can be clearly delineated from the grounds of the Earth. Instead, figure and ground are one; subject and object are inseparable. Context—Earth—comprises us.

In this volume, we invite readers to imagine themselves as permeable co-beings whose existence extends from the collaborative effort of generations. We ask readers to consider: What kind of relationship with Earth have you inherited? What patterns, movements, and modes of attention have you learned from your parents and grandparents, from the places and stories that formed you? These questions remind us to move through our lives with intention, with an awareness of what ought to be, what ought to be left behind, and what kinds of repair we ought to enact in our thoughts, deeds, and words.

The essays in this volume accomplish just that. Each author folds an array of interconnected relations, musings, calls to action, and invitations into their individual experiences of Earth. They reflect upon Earth's many voices, eyes, and mirrors. Through these essays and poems, readers can glimpse and enter these distinctive worlds and ponder pressing questions, insights, and revelations about special moments that have brought our contributors greater awareness and appreciation of Earth.

This volume shares stories from multiple perspectives. We walk alongside those who engage in land-based education and healing, learning from and with young people and children who draw wisdom from Earth. We read about how connecting to and working with Earth can help intergenerational relatives tend to histories of oppression and ultimately find solace in Earth's gifts. We witness the presence of Earth's regenerative capacity to heal, connect, and provide sustenance on tribal lands, in urban gardens, in deep layers of rock, ocean, and soil.

The contributors to this volume contemplate mortality and durability, along with the persistence of monuments and structures that reflect Earth relations and processes. We are invited to think about the impacts of artificial light and other forms of pollution, and the ongoing toll of large-scale extraction that takes from the Earth through violent methods with severe interspecies consequences.

We pivot between existential anguish and elemental joy within the pages of *Earth*. There is a clear naming of pain and disturbance as authors describe conditions of Earth's destruction and related suffering in current human societies. There are also beautiful, sweet moments that reveal the authors' insights, revelations from Earth that offer hope through complex memories and stories. This volume is a treasure that weaves together eclectic experiences. Each contribution brings an expression of how to live responsibly on, with, and through Earth.

Please join us in celebrating these authors. We hope you will find in their stories a path toward an ethic of interdependence, connection, and care for planet, place, and relations. Welcome to *Earth*.

A Garden of Earthly Delights: Bearing Fruit in the City

Hannah Eisler Burnett

I n the summer of 2021, I was six months pregnant and standing on my tiptoes on a footstool, trying to reach an especially ripe cluster of cherries in the tree on my corner in New York City. The cherries ranged from bright crimson to a deep red that was almost black. Those dark cherries were the sweetest, but most of them had been gleaned from the low branches by other urban foragers before my arrival. So I carried my kitchen stool out to the tree, reaching my hands high above my head to grasp at the ripe fruit. After collecting enough to fill a small mixing bowl, my palms were stained mauve, and my fingers were sticky. Looking up, I noted that there were still hundreds of cherries on the branches above me. Leaving them for the birds, I held the bowl close, folded the stool, and walked home through the early summer sunshine full with a feeling of welcoming abundance.

Ripeness is perhaps an overly common metaphor for pregnancy. Still, I came to value my relationship with the fruit trees in my neighborhood not only for their ability to nourish me in summer but also for reflecting back to me the stresses and joys of pregnancy—of bearing fruit—in New York City while experiencing the changing realities of climate catastrophe in real time. These trees are, to me, a conduit for understanding how to live in responsible relation—and in *aware* relation—with local ecologies and global processes, with place-based toxins and new kinds of heat, and with the abundance that continues to accompany all of these things.

Ingesting Place

The summer I spent foraging cherries while pregnant was my first in Astoria, Queens, and it was the second summer of the COVID-19 pandemic. Like many New Yorkers, I experienced those long pandemic days as a time when meandering walks felt like the safest way to meet my neighbors and begin to feel rooted in a new place.[1] On these rambles through the neighborhood, I met others who were also in pursuit of fruit and community: Eastern European women gathered mulberries and made room for me to harvest in their midst, telling me about blueberries nearby that would be ripe soon. Strangers stopped to help me pull down those hard-to-reach cherries on the corner and compare notes about other local foraging spots. Once, I spied a gardener quietly and carefully tending to newly planted fig and peach trees hidden behind a fence, in an area owned by the city on the banks of the East River.

As I walked, I would name the trees I passed, and I imagined sharing that knowledge with my child. Like me, these trees were newcomers to this place; none was a native species. I wasn't born here, and although many of my father's relatives grew up in New York City, I have neither family stories about the history of this place and ecology nor embodied knowledge about how to live here: it's a surprise when acorns start to drop, the rain transforms to sleet, or the tide swells especially high. Despite their recent arrival, all the trees seemed to be thriving, and I felt a camaraderie with them as I enjoyed their shade, breathed the air below them, and sampled their fruit. I tasted the cherries from my corner, but also crab apples, figs, mulberries, and grapes. I began to notice the shape of the leaves on a tree or a vine, the color of cloudlike clusters of flowers, and to associate the onset of summer with the emergence of a delicious harvest. I held the knowledge close, hoping it would help me make this place my home and imagining that the trees were part of the place-based community my child would be born into.

Astoria is a section of Queens that feels simultaneously urban and not: it is full of intersecting highways, power plants, parks, newly constructed condominiums, shorelines, and tiny duplexes. It is where New York City's massive Robert F. Kennedy Bridge makes its vibrating, graceful descent into the Brooklyn-Queens Expressway, and the elevated subway rumbles overhead across the interstate. LaGuardia Airport isn't too far away, and express buses and airplanes add to a general sense of perpetual motion. All this commotion makes the greenness of the neighborhood a welcome surprise; streets are lined with towering white oaks and London plane trees, and on my walks, I saw fruit everywhere. In a raised yard enclosed with a tall white fence, a young lilac grew alongside a spindly peach tree. A few blocks away, a deep driveway was lined in peach trees whose fruit appeared ripe and ready one day and had all been harvested the next. Farther east, in a narrow side yard behind a chain-link fence, I discovered four potted banana trees— banana trees!—that seemed absurdly healthy in New York City's newly tropical climate. On every block, there were at least a half dozen grape vines, some well kept and trellised neatly and some left to writhe across facades and fences, extending into neighboring parks and gardens.

Two years after my pandemic harvest, in the summer of 2023, the same cherry tree produced almost no fruit. The dearth of cherries was, for me, cause for despair: it was yet another signal of Earth system collapse amid a season of extreme rain, strange temperatures, and choking wildfire smoke that flowed through New York City from Canada. The welcoming abundance I'd experienced in my first summer had led me to expect Astoria to be full of fruit at predictable times: mulberries, peaches, nectarines, apples, serviceberries, and figs all seemed to thrive here, even alongside the chaotic rush of city life. So, when an unusually warm spell in early spring meant the cherry trees near my home burst into bloom months before they normally would, I was worried. Over the course of three eighty-degree days, local trees released so much pollen that the color of the street changed: a fine chartreuse dust coated the gray cement. Despite the abundance of arboreal flowers, no insects buzzed to greet them. The timing of the bloom didn't match the spring hatch of bees and other pollinators. When temperatures plunged shortly afterward, frostbitten blooms littered the sidewalk, and I mourned. The lack of cherries in July and August was not a surprise but yet another revelation of the cumulative, local impact of the unpredictable weather caused by global environmental change.

A Host of Relations

During pregnancy, I craved fruit. That's not surprising; while gestating, a human parent carries much more liquid than at any other time in the life cycle. The quantity of blood flowing through a pregnant person's body increases by 50 percent, and this is one factor that leads to the specific joy in eating and drinking that some label "cravings." Water is sweet, and fruit feels like a gift beyond measure. Hydrating foods were especially delicious to me during this time, and I feasted on watermelon, oranges, and grapes from the grocery store; plums, strawberries, and nectarines from the farmers market; and cherries from the tree on the corner.

As my middle swelled, I was also intensely aware of the fact that my body was transforming these fruits into the flesh of a new being, into the possibility of life. Seeing myself expand with this possibility made it especially evident that the life growing within me would be made out of everything with which I had nourished myself: the food I ate, the water I drank, the air I breathed, and my experiences of stress or ease. In other words, my pregnant body was host to an intersection of earthly entities that were specific to the time and places through which I moved.

I thought of this as I rationed my sampling of neighborhood fruit, which I knew carried toxins as well as nutrients. These trees were fed by waterways below the sidewalk that flowed with runoff from the street near the infamously polluted East River. They grew in soil I couldn't see but suspected heavily comprised the detritus of construction and demolition, like so much of the dirt in New York City. While I was pregnant, I attempted harm reduction; I filtered my tap water and ate organic food. Yet I breathed the same air as the trees, dusted with particulates from car exhaust and faraway wildfires. And I ate—sparingly—of the fruit I found on rambling walks, treasuring the way that foraging pulled my attention to the plants and ecosystems around me, teaching me about the rhythms and natural qualities of this place.

Anthropologist Andrea Ford says that "the material relations of childbearing are an interdependence that has no distance." She calls this intersection of responsibilities and vulnerabilities inherent in reproducing human life "co-being."[2] I was unavoidably, and increasingly obviously, gathering relations into myself. I envisioned the expansion of my flesh as a mode of accommodating this generative knot, my body as a host for this convening of generative—and sometimes degenerative—possibility.

My flesh brought to this gathering its own generational history. As with all mammals, the specific possibility of my life began with the genesis of an ovum in my mother's body when she was still a fetus carried in my grandmother's womb. My maternal grandmother grew up white in rural Mississippi, about thirty miles from the chemical plant that employed her brother, who worked to manufacture the fertilizers needed to maintain an agricultural order demanded by plantation capital. My mother also grew up in Mississippi—as did I. She once told me that she and her siblings liked to chase the billowing clouds that tractors left in their wake as they distributed insecticides to kill mosquitoes across the landscape of her youth. They watched crop dusters fly over agricultural fields from the car window on drives through the delta, they ate tomatoes and beans from the garden, and they swam in muddy creeks and ponds near big farms. These are snapshots of what is known to me about the lived places, experiences, and molecules formative of the seed that became my body. The fetus growing within me would be the materialization of those histories and of living here, now.

My child was born that winter, and he is a manifestation of my body and of everything that has sustained me: the emergence of these relations imbued with his own unique volition. I think of this often, that we Earth beings reproduce not through isolated acts of procreation but by drawing our surroundings—local and planetary, past and present—into a distinct material relation that walks on into the future. These relations are nourishment and harm bundled together: just like a human parent—or a fruit tree—the Earth that I have inherited from the relatively recent extractive regimes of capital, colonization, and cruelty can give only from what it has, can show up only as it is.

Life Begets Life

Fruit is a material manifestation of relationships between place, ecology, and change. It is the result of global processes intersecting with local contexts. The emergence of a ripe cherry requires a plant to manifest and open up a flower at the exact moment a pollinator might fly by and be lured by its specific shape, color, and scent. The plant knows when the time is right because of the temperature, the angle and duration of sunlight it receives, and how much water has recently fallen. Fruit feeds many throughout the cycle of its emergence. In its juiciest moment, ripened by planetary relations with sun and rain, a fruit promises to extend this convergence of energies in a new form. This set of convergences and coadaptations are, to me, miraculous.

I imagine fruit as a kind of contract, a sacrifice and a promise at once: a sweet bundle of nutrients in exchange for a seed's future. This exchange happens within a plant's own body—an investment of energy into the opening of a flower, into fruit flesh and seed—and between the plant and the bodies who receive its pollen, nectar, and fruit. I eat a cherry, I receive its hydrating nutrition, and perhaps I carry its pit with me and plant it elsewhere. A life for a life, life begets life.

If fruit is a contract, what are the terms? Robin Wall Kimmerer has written about the radical abundance of serviceberries, the gratitude called forth by their gifts, and the question of reciprocity. "What could I give these plants in return for their generosity?" she asks.[3]

I was trained as an anthropologist, and I can't resist the observation that a gift, like a contract, exists only in relation; in fact, a gift may create a relationship where previously there was none, and it certainly can shape and extend relationships through time and space.[4] Gifts implicate other beings, entities capable of response. Gifts make the giver visible, knowable, memorable. They suggest exchange, although the terms do not need to be the same: perhaps this is what Kimmerer names *reciprocity*. An exchange of gifts does not need to be material: I eat from local fruit trees, and in return, I notice. I yearn to find more, I look up more consistently, and I observe branches heavy with too much summer rain and leaves browning from dry heat, or flowers abuzz with bees and butterflies and the early, green shape of a peach's first weeks on the limb. I am changed; I become a tree's relation.

The gifts wrapped up in fruit are intergenerational processes, and these processes are an inheritance: a set of myriad earthly interdependencies for living and for passing living on. This inheritance manifests in a plant yielding fruit for its own open-ended future, in the possibility that germinates from the continuation of life, of living. Fruit contributes to the nourishment of others—like me, a pregnant human person who rejoiced in encountering a sweet, hydrating snack in the summer of 2021. When I understand fruit as the possibility for life in a specific place, opened up by the intersection of climate, soil, weather, and care, I become aware of the infinite ways I am implicated in this process.

Writing about the constraints of the modern institution of motherhood, Adrienne Rich notes: "What is astonishing, what can give us enormous hope and belief in a future... is all that we have managed to salvage of ourselves, for our children, even within the destructiveness of the institution: the tenderness, the passion, the trust in our instincts, the evocation of a courage we did not know we owned, the detailed apprehension of another human existence, the full realization of the cost and precariousness of life."[5]

I think of this as an observation that can extend beyond the experience of gestational parenthood and that may also describe the state of ecological relations with regard to the constraints of extractive capitalism. Earth is depleted. Perhaps the unseasonality and careening temperatures of 2023 meant that local trees couldn't offer their habitual gifts. Yet I remain in relation with the trees, compelled by "the detailed apprehension of another" and "the full realization of the cost and precariousness of life." I can pass this relation on, a gift, to my child. Perhaps this gesture meets the terms of a contract I accepted when a cherry passed my lips one bright June morning two years ago.

A Forest for the Trees

In the morning, I hear a song lilting through the closed door of my child's room. I peer inside and am greeted by a smile and, as he stands to greet me, he repeats a daily exhortation: "Look at the forest!"

He thinks the trees out his window are a forest. At first I felt worried by this association: what sort of city creature am I raising, who thinks a paltry six oaks might make up something as abundant as a forest? Yet the trees outside are larger than our building, and at least as old. They dance in the spring wind and drop acorns in the fall. Squirrels run and leap between them, and birds call out from their depths. Someone floors below us keeps bees in hives that are difficult to spot among the garden beds and dappled shade. Someone else has carefully grown a magnificent fig tree, whose branches crown just below our windows.

Pregnancy pulled me into embodied and attentive relation with the trees closest to me: gestation called forth a great thirst, which was answered by the fruit dripping from the branches of my neighborhood cherry tree. Those cherries spurred my attention, and I noticed even more abundant sweetness being offered by the plants around me. My attention to these relationships continues years after giving birth: In the catastrophic summer of 2023, I observed early blooms and untimely fruit drops, the heavy leaves and swollen limbs of local trees that began to collapse under the weight of too much water. It is an attitude of noticing, an awareness of quiet Earth relations that could easily be swallowed by the urgencies of traffic, work, and domestic labor. This attention might be a kind of ethic, an orientation toward life on Earth. We live among beings who share their rhythms, stories, and abundance with us— our shared conditions of possibility. One way to return this gift is to participate in the exchange: to taste the fruit and to feel implicated in its existence—to keep our senses open to entry points into responsible relations with one another and with Earth.

notes

1. See, e.g., Jordan Kisner, "The Lockdown Showed How the Economy Exploits Women: She Already Knew," *New York Times Magazine,* February 17, 2021, https://www.nytimes.com/2021/02/17/magazine/waged-housework.html.
2. Andrea Ford, "Purity Is Not the Point: Chemical Toxicity, Childbearing, and Consumer Politics as Care," *Catalyst* 6, no. 1 (2020): 9.
3. Robin Kimmerer, "The Serviceberry: An Economy of Abundance," *Emergence Magazine,* October 26, 2022, https://emergencemagazine.org/essay/the-serviceberry/.
4. See, e.g., Marilyn Strathern, *The Gender of the Gift: Problems with Women and Problems with Society in Melanesia* (Berkeley: University of California Press, 1988).
5. Adrienne Rich, *Of Woman Born: Motherhood as Experience and Institution* (New York: W. W. Norton and Co., 1986), 280.

Ancestral Mothers of the Hood

Oyah Beverly A. Reed Scott

O yah is my name. My name means "Momma Earth." I am a woman sustained by roots, reclaiming my ancestral, mythical, and literal roots. I am reclaiming sacred stories of my lineage as a Black woman descended from native people on three continents. In their memories, I find meaning and place.

The staples grown by my maternal grandmother on an allotted patch of Earth in Sunflower County, Mississippi, were essential to the survival of her eleven children as they struggled to make ends meet in the rural South. The same is true of my paternal grandmother, who had just one child, my father. My grandmother worked the land and also took in ironing. She and my dad spent his early years growing and going to market with the staple vegetables of Louisiana soul food: okra, onions, celery, greens, and tomatoes.

Although these two women were in different states, their circumstances kept them close to the earth. It is their faith, strong sense of service, and desire to see and be more that was embedded in the character of my parents, who met, married, and in 1962 settled in the Englewood community of Chicago to build a life together. It is ingrained in my spirit to serve my community through my connection with the earth.

Miss Mary and the Green Team

Sometime in April 2000, a wonderful woman and neighbor of mine in Chicago, Miss Mary, died unexpectedly. She was the mother of my son's good friends. They lived around the corner from where I had grown up and had returned to raise my older children. As was expected, when she died, I went to visit the boys, family members, and neighbors. The young men on the block had spent a great deal of time at Miss Mary's and they were clearly upset but also seemed lost and unmoored.

When I saw their distress, an overwhelming wave of grief and intuition came over me. I had the strong sense that I should create some kind of memorial on a parcel of vacant land to honor Miss Mary and the other mothers who had left strong impressions on our community. I could feel these mothers' affirmative ancestral energy as I pursued and secured funding to create a garden and other opportunities for young adults for whom college was not an option.

When Miss Mary passed away, the Ancestral Mothers chose me to pass my ancestral qualities on to the neighborhood's next generation. Several serendipitous events occurred. There was a grant opportunity on the table for a community garden. Walmart was seeking a foothold in the Black community as a solution to the food deserts that poor neighborhoods endured, and a vacant lot was available. Armed with a grant, with Walmart as a sponsor, and with permission to use the land, I called a meeting at my dad's house and our Green Team was born.

Gardening: A Sacred Science

Technically, soil is a mixture of minerals, dead and living organisms, air, and water. These four ingredients react with one another in amazing ways, making soil one of our planet's most important natural resources. There are three types of soil. These groups are determined by size and identified as sand, silt, and clay. An equal mixture of the three types is called loam. Soil structure is the arrangement of soil particles into small clumps called peds. A soil's texture and structure tell us a lot about how soil will behave. Granular soils with a loamy texture make the best farmland, for example, because they hold water and nutrients well. Single-grained soils with a sandy texture don't make good farmland because water drains from them too fast. This combination of science and the sacred would matter greatly as I undertook the request of the Ancestral Mothers.

My goal in leading the Green Team was to reawaken curiosity in and knowledge of the earth. This neighborhood had meant the promise of a greater future for my parents, and I wanted to grow that meaning in the young people. I could see Dad and the other men of the neighborhood gathered in our backyard discussing what would be planted that summer. They bragged about the quality of the soil and how good they were as gardeners. The men on the block had come up from different parts of the South and brought the knowledge of how to live on this earth with them. How to plant what was good for both the farmer and the soil—they knew that. When to plant what and what to plant with it.

The southern portion of the United States was present in that yard, holding that soil in hand. Mr. Giles came from Alabama, and he insisted on cabbage and peas. Mr. Webber was a son of Mississippi; he preferred okra and onions. Mr. Ham of Arkansas didn't have a preference. He planted whatever he wanted but mostly tomatoes and collard greens.

My daddy had a green thumb. Everything he planted did well, usually tomatoes, okra, bell peppers, collard greens, and one memorable year he grew sweet corn. That year, my mother made her succotash with all the vegetables from the garden, to go along with the fried chicken and white rice. These meals anchored me to the earth. I felt the same respect, awe, and love for the earth as I did for my beloved great-grandmothers.

I had no choice; I am the daughter of workers of the land. I am the combination of Louisiana and Mississippi workers whose lives depended on what they could grow. I learned early how to walk barefoot on the soil when I needed to heal my heart. I knew to put my hands on the earth when I needed to scream but couldn't speak.

The earth beneath my feet has kept me from leaving the planet in times of despair, disgrace, or fear. Momma Earth absorbs, warms, heals, holds, and hides us away. I prayed that everyone who participated and everyone who walked or drove past the Green Team's Garden Project would feel a portion of their soul soothed.

Our Learning Community

I developed a program that included two days a week in the classroom. Our office was located at Northeastern Illinois University in the historic Bronzeville community of Chicago. For most of the group, it would be their first time being in a college environment. These were young people who had been handed unseemly roles; it was as if for much of their lives they had been placed in scenes and forced to perform material written by plagiarists. Spending time in the classroom was an important part of our work together: they loved occupying that space as accepted members of the learning community.

Unlike other parts of our lives, our garden vegetables were never meant to harm us. Our soul garden was meant to contain grief and rage in those raised beds along with the okra, onions, greens, and tomatoes. We preserved the memory and dignity of the mothers, grandmothers, and young brothers who died from society's pesticides—callous neglect, poverty, and racial divides. I taught them to have pride in the gardens of their grandmothers and pawpaws, to experience the mystical synergy of Black hands in Black soil. I loved to witness their seeds of memory unfurl themselves from the subconscious and extend their ancient stories into these beautiful humans' hardened hearts.

The Soul Garden

Every function of gardening they learned had its own alchemical purpose for their souls. The history of the vegetables we planted, the history of the land we visited and the land we planted on, the benefits of growing one's own food, the types of watering sources, of planting beds, and climate lent themselves to the stories I would tell about their lives. They were required to take notes, and after listening to lectures, viewing films, or listening to music, we would engage in robust conversations.

Nearly all the information was new to them. They were very curious about the subject matter and asked interesting and often-times humorous questions. For example, they asked about the difference between soil and dirt. I told them soil is what dirt wishes it was: enlivened by its friends and family—that is, nutrients, microbes, useful insects, organic matter, even manure, because we all have to deal with some BS sometimes; it's what we do with it that makes all the difference.

Soil shows us what to do with the cards we get dealt in life. In many ways, Black youth are treated as if they are the depleted version of "real" young people. And like depleted soil, a depleted soul is unlikely to produce anything of merit that could take root and bear fruit. One day I thought to myself, *The ancestors have called me to provide the compost for the souls of these young men.* In that moment I felt as if I had opened a locked door and found a hidden treasure. I am the compost; I help give them access to the earth in a way they never imagined. I can show them the mystery and magic of the garden.

Water

There are many ways to describe the process of fertilization. When water is applied to the soil, it seeps down through the root zone very gradually. Each layer of soil must be filled to "field capacity" before water descends to the next layer. As it is with water, so it is with wounded humans. Love and purpose acted on and in the young people like water entering the soil. Water moves downward through a sandy, coarse soil much faster than through a fine-textured soil such as clay or silt. These men of clay and silt gradually began to receive respite from the daily grind that the project offered, and the result was not miraculous but it was real—and it was intractable.

The earth herself was the greatest teacher. The vacant lot had once been a nameless space for trash and tragedy, but not anymore. Their direct role in transforming the land had a real effect on the young people. As they practiced the principles of sustainable living, I began to see a change in the way they carried themselves. It was a modern-day miracle of them all healing their wounds and flourishing. The load had lessened on the lives of these young people. They were rarely late for the garden or the classroom. They had turned the soil and felt the power of a hard day's work on the land. We prepared healthy lunches at my dad's house and ate them in his backyard. There was easy laughter and plans made for the land dedication. They participated in local Chicago parades. The land dedication was like the prom many of them had never attended.

Seed Saving

Seed saving was another way they got in touch with the natural world, a practice through which they felt they were a part of something greater than themselves. After three hundred generations of humanity saving and sharing seeds, in just three generations we have stopped these practices and placed ourselves in the position of potentially losing 90 percent of our seed diversity. Local people offered seed-school training as a way to increase interest and knowledge of basic seed-saving techniques, to encourage the reclamation of the time-honored and, I would say, sacred role of seed saving.

The environment lent itself to informal but well-organized training. We learned basic flower and seed classification, including the family, genus, and geographic areas where each species would flourish, as well as techniques for seed saving and how to test the viability of seeds using simple methods such as laying seed on a damp paper towel to check for sprouting. We learned how to re-purpose items, like using a kitchen strainer to salvage seeds from cantaloupe.

A Harvest of Hearts

One season, we hosted a land dedication with activities that included libations and ancestor calls, when the audience was invited to call out the names of their own loved ones who had transitioned to the next life. For each name, we added an addendum to the roll call. In the planning of the event, we discussed the ways we could demonstrate our knowledge, our creativity, and our culture. We demonstrated the bridge they were building across time and space. The land dedication included a luncheon at the garden site, a dedication, ancestor call, ceremony, performances, and speakers.

Our garden was born out of the idea of commemorating the lives of the Mothers who had made their transition, although we also added the members of the community who we had lost over the summer to gun violence as well. We commemorated the ancestors by designating Jay to be the "keeper of the names."

During special events, like when we got our second grant, when we held our land dedication, and the closing of the summer garden, we would retrieve the scroll and call out the names of the dead as we poured libations. The woman who administered the grant for the state made a mandatory visit to the garden during the summer. The group members were in rare form as they displayed their knowledge of gardening and the importance of creating their community garden. They explained the name of the garden. They explained their connection to the land and to the community, from the Delta to the neighborhood. At the end of the meeting, she posed for a photo, with quite a bit of joy, and participated in the call-and-response "Green Team/For Life."

The garden served as a place of remembrance, celebration, education, and community. Nearly all the families in the neighborhood had come up to Chicago during the Great Migration, and the vegetables we planted honored their journey. Okra, onions, greens, and tomatoes were foundational foods for our community. By planting them, we could connect our current experiences and hopes for the future with our past.

We invited the community and guests to call out the names of loved ones who were currently serving time in one of America's many privatized prisons as we sang Bob Marley's sweet liberation song, "Three Little Birds." With these experiences, I feel as though I answered the call of our neighborhood's Ancestral Mothers. I was to create a healing opportunity for the young people of the neighborhood, a call driven by the untimely death of Miss Mary. I did that. I read, researched, visited sites, went to school, became a master composter, negotiated for land, wrote grants, bartered for goods and services, acquired sponsors, taught, provided civic lessons, provided lunch and a check, and loved the young people as my own children.

Growing with Life

The garden is long grown over now. Nearly fifteen years have passed since Miss Mary died. The Green Team has seen its share of heartbreak and disappointment. Miss Mary's children moved out of state, but both are thriving. Other members, like Alex and Mook, were murdered. A couple Green Team members served time but have bounced back and are doing well. Some disappeared, and other members suffer from heartbreak, illnesses, or alcoholism but still maintain their careers and families. Several men went on to have good careers in areas related to the green economy. My dad is now ninety-six and still planting a few tomatoes in the backyard.

By murder or accident, we have lost four members. Several left the neighborhood, and most have done something positive with their lives and are raising families, paying taxes, and voting. They still get together on the block, usually behind my dad's house, playing music, shooting the breeze, and pouring libations before swallowing their grief. Nowadays when they see me, they all take turns coming up to me for their hug and they still call me Momma Earth. Sometimes, in my overwhelming love for them, I call out: "Green Team!" And they respond, "For Life!"

postscript

This essay is dedicated to the memory of Alex Fleming, a Green Team member who became a son to me. Alex was my biological son's best friend, and our relationship was close: he called me Mom. As I was writing this essay, I got a call telling me he had been shot. When my son Michael came to the phone to speak to his lifelong best friend, I couldn't understand him. His grief was vast—as wide and deep as the Atlantic Ocean—and in his wailing, I heard the cries of the ancestors. I thought about the numbers of the righteous dead that lay in the body of Mother Earth. How, in her wisdom, she transmutes our bodies and how, in our ignorance, we fight the inevitable with concrete vaults and caskets. I helped Alex's mother plan and conduct his funeral. As I stood at the podium to give Alex's eulogy, I looked out at the crowded room and saw most of the Green Team members were present. After they had lowered Alex into the Earth, we found we were the last ones standing at his grave site. Softly we sang, through tears and with our precious memories: "Okra Onions Greens Tomatoes, 7031 S Sangamon." I gave the call: "Alexander Fleming, King of the Bud Billiken Green Team. The King is dead. Long live the King. Green Team. For Life." Please visit YouTube (https://youtu.be/-Tm9FONf8oM) to hear the Green Team singing our anthem.

the sky is made of dirt

Alexis Pauline Gumbs

At some point, someone noticed all the mica in the sky,
dirty and bright beside the light of the moon. It would be
nice, one grandmother said, if we could work with clay
like that, imagine how our bowls would shine. It would be
beautiful, another grandmother said to pour water from
a vessel iridescent like that. Like drinking stars, her sister
smiled. That was nice. They continued with their work. No
one had time to run off to the sky to get different dirt. But the
grandmothers laughed and dreamed. And the little rascal
granddaughter took note of their stretching mouths, their
glinting eyes, fluttering hands.

Someday people would call the person the little rascal
granddaughter became a holy one, a gatekeeper, a shaman, a
seer. Someday people would reach back and try to remember
the songs, the dances, the riddles, the ceremonies this type
of trickster did. But at the time, all we knew was that this one
wasn't going this way and wasn't going that way. This one
wasn't following the mother or following the father, but also
both? This one was always getting in the way. Saying listen
to this. Guess what I saw? Tell me your dream! Give me your
face! When we let her, she smudged her sweet fingerprints all
over our skin.

Maybe she whispered it to us while we slept, home from
her nights looking up at the stars. Maybe she taught it to the

children while she showed them where to find the crabs and how to greet the water. Maybe we just gradually noticed that there was something about the mud around us that was moving us to something more. Maybe we all already knew the craving under our tongues and just needed to finally admit it. But when that storm came and took so much of the land that the roots of the trees couldn't hold it, we all remembered. We asked each other, *what about that glitter dirt in the sky?*

And of course here came our rascal, smiling like a million grandmothers, hands fluttering in revelation, eyes shining like *oh my beloved people, I thought you'd never ask!*

Here was the plan, she drew it in sand with an awara palm leaf in her hand. Here is the plan, she waited until right before sunrise. Here is the way, she drew it in swirls right on the edge of day. Here it is! She declared and behold what she drew in the sand matched exactly the sky!

And who will go fetch the clay? Someone asked. And the mothers all kissed our own teeth. We didn't have to turn around to see what we must have been waiting for anyway. All our craziest daughters. Raising their hands.

And so it began.

The Chronicles of Earthsea

Marcia Bjornerud

Although Western science long ago abandoned the classical elements for the crowd on the modern periodic table, the original four still loom large in the way many people conceive of the world. The prepositions we use with the primal elements reveal much about our relationship with the realms they define: To be "on fire," "in the air," and "at sea" are all states of disequilibrium and transition, while being "down to earth" signifies stability and permanence—and implicitly equates the soil beneath our feet with the planet as whole.

But our view from terra firma is biased. As the science-fiction writer Arthur C. Clarke is said to have quipped, "How inappropriate to call this planet Earth when it is quite clearly Ocean." Whether or not Clarke actually uttered it, the quote is a good reminder to us landlubbers that our conception of the world is blinkered, and the global environment is actually governed by the realm of the seas. The water cycle is overwhelmingly oceanic; almost 90 percent of annual evaporation and 80 percent of precipitation on Earth takes place over the ocean. The fraction of ocean-derived water vapor that doesn't fall back into the sea is essential to life on land. The enormous heat capacity of the oceans acts to modulate planetary temperature fluctuations, and the mighty global ocean currents work incessantly to shuttle heat from the tropics to the poles, dictating the kinds of ecosystems that can exist above sea level. It's hard to deny the oceanic character of Earth.

As a geologist who studies the solid Earth, however, I can't help pointing out that even the vast Ocean has a floor—and that floor is more Earth. Also, the geological boundary between continents and oceans is less sharp than one might think. In fact, the liminal zone between them is one of the strangest things about this planet, constituting a third, distinct realm of Earth.

Borrowing from another great science-fiction writer, Ursula Le Guin, we might call that realm "Earthsea." Before we can fathom this hybrid domain, however, we need a deeper understanding of the character of the adjoining territories and their complex interdependencies. Earth and Ocean, these elemental entities, are not as distinct as they may seem, overlapping and interacting in surprising ways.

Sea

Astonishingly little was known about the nature of the crust beneath the deep oceans until as recently as the 1950s, when classified data collected by the US Navy during World War II became available to scientists. New maps of submarine topography revealed an alien realm entirely different from the landforms of the continents: a globe-encircling chain of volcanoes—the mid-ocean ridges—flanked by vast abyssal plains bordered in turn by vertiginous trenches. The new data put to rest old ideas that the oceans were underlain by the same kind of rocky crust as the continents but crust that just happened to be covered by water.

Geologists were surprised to discover that unlike the geologically heterogeneous continents, which are built from a wide range of rock types formed over billions of years, the ocean floor was globally uniform—endless expanses of black basalt—and none of it was older than 170 million years. This led to the realization that ocean crust is created, and destroyed, by entirely different processes than those occurring on the continents. Specifically, ocean crust is formed by volcanism at mid-ocean ridges through the phenomenon of seafloor spreading. Then it gradually cools and subsides until it is dense enough to sink back into Earth's interior in the process called subduction. At subduction zones, marked by the terrifying trenches on the seafloor, slabs of cold ocean crust slide beneath an overriding tectonic plate at an average pace of several inches per year—but intermittently much faster. Subducting plates may lurch abruptly downward in great earthquakes that can unleash giant tsunamis, fearsome illustrations of the combined power of Earth and Sea. Although from a human perspective such events may seem only destructive, the practice of subduction—a crustal recycling process unique to Earth—is essential to the long-term habitability of the planet.

Subduction not only returns ocean crust back to the mantle but also carries water contained in those crustal rocks down into Earth's interior. Geophysicists estimate that the rocky mantle contains at least an ocean's worth of water—a precious planetary savings account that ensures there will be Ocean in the future. This great internal reservoir is tapped over time by volcanoes that vent water vapor back to the surface. The interior of the planet is, in a sense, both Earth and Sea. Subduction of ocean crust is, in other words, a way for Earth's atmosphere and hydrosphere to remain in close communication with Earth's interior. No other rocky planet or moon in our solar system has developed this habit. On our neighboring planets, water and other light compounds such as carbon dioxide have simply been released over time through volcanism and then either accumulated in the atmosphere (as on Venus, a runaway greenhouse) or were lost to space (as on Mars, a frigid desert).

Although Venus and Mars don't practice subduction, they are worlds made of basalt, the same rock that forms Earth's ocean crust. In fact, these planets—like artists obsessed with a single color or motif—have never produced anything besides basalt from their mantles. Earth, in contrast, has distilled from a mantle of the same composition two distinct types of crust and invented a huge variety of rock types. The differences in creative output among the three planets can be seen on graphs showing the distribution of their surface elevations. This type of graph is called a "hypso-metric" plot, a Greek neologism meaning "height-measure." Such a plot is essentially like a histogram of exam scores showing the range and distribution of scores earned by members of a class. On a hypsometric plot, the "scores" are elevations on the solid surface of the planet, ranging from lowest to highest, and the "number of students" is the percentage of the planet's surface area that lies at those elevations.

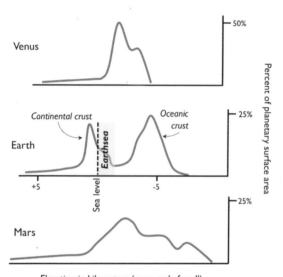

Elevation in kilometers (same scale for all)

Hypsometric plots for Venus, Earth, and Mars showing the distribution of surface elevations. The figure is adapted from S. Mikhail and M. Heap, "Hot Climate Inhibits Volcanism on Venus: Constraints from Rock Deformation Experiments and Argon Isotope Geochemistry," *Physics of Earth and Planetary Interiors* 268 (2017): 18–34, https://doi.org/10.1016/j.pepi.2017.05.007.

It's possible to make hypsometric plots of other planets be-cause decades of space missions—many with names that point back to Earth's oceans (Viking, Mariner, Magellan, Odyssey)—have provided detailed elevation data from our neighbors in space. (In fact, orbiting satellites have yielded higher-resolution topographic maps of the surface of Mars than of Earth, a consequence of all that Ocean here on Earth). When we make hypsometric plots of these other bodies, we see that the absolute range of surface elevations—the differential between the lowest and highest points—varies from 14 kilometers (8.4 miles) for Venus to an impressive 31 kilometers (18.6 miles) for little Mars. The difference between the highest and lowest points on Earth—the top of Mount Everest at 9 kilometers (5.4 miles) above sea level and the depths of the Mariana Trench in the Pacific at 11 kilometers (6.6 miles) below—is a comparatively modest 20 kilometers (12 miles). But the most striking pattern that emerges when these hypsometric plots are seen together is that Venus and Mars have a single "mode," or peak, representing the most common surface elevation.[1] Earth's plot, in contrast, has two distinct peaks, reflecting the existence of two distinct types of crustal material: one (basaltic oceanic crust) that is low and dense and one (granitic continental crust) that is high and buoyant, like ships with different kinds of cargo riding at different elevations in the water. In other words, the bottom of Earth's oceans are not simply low-lying land; they are made of fundamentally different stuff than the continents. And conversely, even if Venus and Mars had oceanic quantities of water, they would not truly have distinct continents and ocean basins, only rugged plains of basalt pockmarked by meteorite impacts and piled up here and there in volcanic heaps. In contrast, Earth's ocean basins, the vessels for this planet's profligate volumes of water, are defined by their rocky contrast with the continents.

Earth

If we take this planetary point of view, we see that Earth's continents are the true enigma. Although all ocean basalt on Earth is geologically young—recycled on timescales of a hundred million years or so—basalt as a rock type is considered primitive, or primary, in the sense that it can be derived in just one simple step from the mantle by melting just the lowest-temperature components. All our near neighbors in space have managed that simple recipe. So how does a planet cook up continents?

The answer: slowly and patiently. Unlike the literally monolithic (one-rock) ocean crust with its short, predictable life story, the continents are three-dimensional mosaics of many rock types with idiosyncratic histories formed over long periods of time. If one could blend the rocks of the continents into a homogeneous puree, however, the composition would be broadly granitic. The ancient basement rocks of most continents—their long-lived foundations—are typically granites or similar rocks, light in density, dominated by minerals like milky quartz and pink feldspar. No other rocky planet or moon seems to have produced granite. That is, Earth, a.k.a. Ocean, could also be called Granite. As it happens, all that eponymous water plays a role in making that signature granite.

Water carried by a subducting slab into the mantle gets baked out of the oceanic crust at depths of about 40 kilometers (25 miles). This water then rises into the mantle material above the down-going slab and acts to reduce the melting temperature of the mantle rock, generating magmas of different composition than those produced by seafloor spreading at mid-ocean ridges. Sometimes these magmas, in turn, cause secondary melting of the overlying crust, distilling the lowest-melting temperature constituents from those rocks and forming new magmas that are several compositional generations removed from the ancestral mantle. In this way, little by little, over billions of years, the granitic continents were built.

Once formed, continents are not easily dismantled. The oldest known continental rocks, the Acasta Gneisses of northwestern Canada, are more than four billion years old—more than ten times older than the oldest surviving ocean crust. Granite, even when it is old and cold, never becomes dense enough to be subducted into the mantle. Seafloor spreading may break continents apart, and subduction can cause them to collide and buckle, raising great mountain belts—but these processes do not change the volume of continental rock. Erosion, you might be thinking, is an obvious way to destroy a continent, and it is true that rivers tirelessly tear down highlands, carrying the detritus with them as they race toward the Ocean. But here is where the third realm, Earthsea, the demimonde between Earth and Ocean, joins our story.

The Chronicles of Earthsea

The hypsometric plot for Earth reveals a mismatch between the geological and the geographic definitions of continents and oceans: specifically, the lowest-elevation granitic (continental) crust lies below the height of mean sea level. This region of submerged continental rocks—neither continent nor ocean yet both continent and ocean—is the dominion of Earthsea, known more prosaically as the continental shelves. The continental shelves are areas where the granitic crust has been attenuated by tectonic rifting—usually during the opening of a new ocean basin—and is too thin to stand buoyantly above sea level. This netherworld is a realm of profligate biodiversity, its waters rich in land-derived nutrients and shallow enough to be illuminated by sunlight.

What makes the shelves of Earthsea even more remarkable is that their hybrid character exempts them from the forces that destroy either fully continental or fully oceanic rocks. Because they are geologically continental, they cannot be subducted; and because they lie below sea level, they are not vulnerable to the rapacious erosion of rivers, who lose their energy at the shore and spill their sedimentary freight out onto the shelves, where it accumulates layer by layer. The continental shelves therefore act as faithful archivists of both the terrestrial and the aquatic domains, keeping records of events on land that the eroding continents themselves won't preserve while also chronicling changes in ocean chemistry and marine ecosystems, which the short-lived ocean

crust quickly forgets. In other words, the sediments of the conti-
nental shelves—the chronicles of Earthsea—are our most detailed
account of the atmosphere, hydrosphere, and biosphere over geo-
logical time, recording evolutionary innovations, environmental
upheavals, climate oscillations, and mass extinctions. It's a long
and rambling, fascinating, humbling, and occasionally terrifying
saga. Earthsea holds the history of the world.

Still more remarkable, these continental-shelf archives are
episodically reshelved by tectonics, like a library that periodically
consolidates its volumes into a more compact space. When a long-
lived subduction zone finally closes an ocean and continents col-
lide, their shelf areas, laden with stratified sedimentary treatises,
are the first to buckle and fold, forming mountains. In this way,
those carefully kept chronicles are shoved up into the open air,
where they become accessible to those land-dwelling creatures
who are under the impression that the planet is Earth, not Ocean.

It almost seems that having gone to the trouble of creating
such a record, the authors, Earth and Ocean, want to make sure
that it will be read. This magnum opus could have been written
only through the intimate collaboration of these great elemental
powers. The chronicles of Earthsea challenge Western science's
centuries-old belief that truth will be found by distilling the world
into "pure" elements. The unruly world spills out across our tidy
categories, revealing its richest truths in the hybrid realms betwixt
and between.

notes

1. Mars does have two distinct terrains that lie at different elevations—the ancient,
   heavily cratered southern highlands and the smoother, lower northern lowlands—
   but that contrast probably reflects the origin of the lowlands as a single great-
   impact crater, not two fundamentally different types of crust. See O. Aharonson,
   M. Zuber, and D. Rothman, "Statistics of Mars' Topography from the Mars Orbiter
   Laser Altimeter," *Journal of Geophysical Research: Planets* 106 (2001): 23723–35,
   https://agupubs.onlinelibrary.wiley.com/doi/abs/10.1029/2000JE001403.

Layers of the Turtle's Shell

Laticia McNaughton

Tsi Tewahsénhtha: "The Place Where the Water Falls"

M y family and I visited various places in the Niagara region during the weekends when I was off from school. These excursions usually took place in the summertime, along the Niagara Gorge, the shores of Lake Ontario and Lake Erie, Eighteen Mile Creek, the Iroquois Wildlife Refuge, "The Ridge" on the Tuscarora Nation territory, and other small creeks and tributaries in western New York. My father was passionate about this region that is nestled between two of the Great Lakes and one of the greatest known waterfalls; he always touted its exceptional geological history and wonder.

Human beings are said to have been in this region for only 150,000 years, yet the deepest layers of the Niagara Gorge are estimated to be 448 million years old.[1] The tiny fossils my family and I marveled over during our hikes were once living creatures who were part of an ancient sea dating to that age. This territory was once covered by waters and lakes, most notably Lakes Tonawanda and Iroquois. Mammoth glaciers slowly carved out the landscape, receded, melted, and created the freshwater source of the Great Lakes we know today.[2] The glaciers' movements and their shaping of the terrain made it possible for us to access the exposed fossils.

Created by similar geological processes, the stone ledge and clifflike landforms of the Niagara Escarpment make up a massive limestone-based ridge connecting far northern Quebec to Wisconsin and northern Illinois.[3] This steep ridge and natural boundary create different winds, temperatures, and seasons along southern Lake Ontario. The lakes encourage intense lake-effect snow squalls, and their microclimates influence agricultural pursuits.[4] These great geological features continue to shape and play a generative role in the local climate and daily lives of the people who live here.

The people of the Haudenosaunee Confederacy have long called this ancient, fertile, and unique landscape their home. The Seneca and Tuscarora Nations, in particular, have been caretakers of the land. The Tuscarora Nation arrived later, continuing a vast agricultural legacy reminiscent of their North Carolina homelands. Before the Seneca Nation and Tuscarora Nations, the Tyonontahte, Attiwendaron, and other nations took up residence in the region. From hundreds of years of observation, traditional teachings, and ecological knowledge, our ancestors came to know that this land is well suited for its agricultural possibility—limestone-rich soils, climate, and even spiritual power. This history reminds us of the importance of place-based knowledge, connection, good stewardship, and continued responsibility to the lands upon which we live.

Imprints

My family often trekked across the rocky, elevated terrain that is the Niagara Gorge. The trails were steep and slightly treacherous as we climbed down off the main paths, holding onto slanted trees to steady ourselves. The lower we descended, closer to the rushing Niagara River banks, the more we sensed the age of the stone walls growing older.

The giant stone towered over me—to the left and right—as though I were in a canyon. Well, I *was* in a canyon of sorts. Above us, hundreds of feet of the gray and brown sedimentary rocks loomed, their layers carved out by the glaciers in this region so many years ago, which manifested in thousands of parallel lines and geometrical patterns. They are a monument of deep time passed.

My father and I would carefully examine the wet, flat slabs of stone. We inspected them closely, and the imprints of tiny shells and ridges appeared as ghostly echoes of fossilized aquatic critters and plant life from the past. Brachiopods. Corals. Crinoids. Trilobites. Sometimes, when we were lucky, we found a whole specimen embedded within the ancient layers. The tiny sea creatures who once roamed those waters thousands of years ago reminded us of our small part in the earth's life and its processes.

With all the rocks surrounding us, my father, sister, and I were reminded of the soil and stone upon the Turtle's back. In the Haudenosaunee Creation narrative, the turtle generously gave us the space upon which we live. When I reflected on this knowledge, my teenage self glowed with curiosity about the earth and its wisdom. In the Niagara Gorge, under the gaze of great slabs of stone, I was simply far beyond the mundane adolescent worry of chemistry exams.

Once we reached the shoreline, I gazed to the left and right, looking up on the land that on either side is called Canada and the United States. It's really called Tsi Tewahsénhtha in the Mohawk language, the language of my people, which translates to "the place where the water falls" or "the place where a great stream is falling."

These experiences with my family led to questions: Are we humans but a blip in seemingly infinite time and history? Yet we are part of this ecosystem too—we belong with the waters, sky, and land. Being in the presence of the earth's power invites us to ask, What does it mean to be human? What gives our lives meaning? What is our role in this vast system? What can we learn during our brief journey in this much younger, present-day geological age? How can we be good stewards of the earth and cultivate a relationship within a vast network of plants, trees, fungi, fauna, rocks, and old earth?

Regional land survey reports dating back to the mid-nineteenth century describe ideal conditions regarding the terrain, climate, and soil for American settlers who were interested in timber and agriculture.[5] Haudenosaunee people already had an appreciation for these conditions, along with a vast and sophisticated infrastructure of agriculture and food security. Their agricultural operations were well documented by soldiers' accounts from the late eighteenth-century Sullivan Expedition. A euphemism, the "expedition" was a tactic of war, systematic and genocidal, ordered by General George Washington in 1779 to "ruin their crops now in the ground and prevent them from planting more."[6]

Teachings from the Stones

In addition to the Niagara Gorge and Niagara Falls, I also hold a deep love for my grandmother's land on the Tuscarora Nation territory. As a youth, I was one with that earth, playing with mud caked on my skin, picking black raspberries in the sweltering July fields, growing corn in its rich soil, diving into autumn leaf piles, or tracking animal prints in the snow. I grew fond of the woods that lay at the back of my family's land. I found comfort and solace in a grove of giant mossy boulders in that lush forest. The boulders sat there mostly undisturbed, in clusters, under a great canopy of trees, with barely any sun peeking through. I walked there in all seasons, meditating on the stories the stones held, the people they'd seen or met, the animals who came to appreciate them, or the insects who took shelter within their dark crevices. I could imagine the boulders watching the cycles, season after season, of humus and mycelial networks as the earth in this part of Turtle Island freezes, thaws, swelters, steams, wilts, and goes to rest yet again. Those boulders hold thousands and thousands of years of memory. Thousands of winters they've seen and survived, stationary yet also moving alongside us as we rotate around the sun.

Rocks are conventionally understood as an inanimate part of the landscape, as quiet background figures. Yet the stones, rocks, and layers of sediment are so much older than us. Rocks figure greatly into Haudenosaunee teachings and lore. They carry with them a knowledge based on the memory of hundreds of thousands of years.

I'm reminded of a Seneca story about a stone who was seen as an elder figure.[7] In this narrative, a young man with no family was cast out of his village. Wandering alone, he befriended a talking stone who shared stories and knowledge about hunting in exchange for a tobacco gift. Every day, the stone would impart its teachings and the young man would leave a gift. Soon, the young man carried the teachings back to his village and shared the gift with them, restoring his place in his community.

The Haudenosaunee Creation narrative is a grand epic that includes Haudenosaunee foundational teachings and philosophies about the origin of the world and human life. A pair of twin brothers, Okwirá:seh and Tawískaron, held special abilities to create different elements, plants, foods, animals, and natural features on the earth.[8] Okwirá:seh, or Sapling, especially created many things, exclaiming: "I tell you that the earth is alive! Now you must understand that I took earth and from it I created all these things that live."[9] Part of his creations included the rivers and stone mountains, the geological features of the earth's landscape. Sapling came to be acknowledged as an important being for his ability to both create and move mountains. Holding this ability to shift the earth's landscape meant the brother held great power and was to be respected. Sapling was also responsible for the creation of humankind.

The other twin brother, Tawískaron, whose name can be interpreted as "flint" or "stone," also held creative powers to make things, but the features he created were undesirable, harmful, or poisonous. Flint, jealous of his brother, sabotaged Sapling's creations by hiding the animals Sapling created inside a cave covered by stone. This later led to a dispute as the twins came into conflict and decided to settle their differences with a game.

The Genohskwa are another example of how stone and rock feature prominently in a number of Haudenosaunee stories. The Genohskwa, also called the "stone giants" or "stone coats," were beings who had skin made of rock. Formidable creatures with great powers and abilities, their appearance in the tellings often meant the demise of whomever they encountered.

Even the Peacemaker in the Great Law of Peace narrative built a stone canoe and steered it over treacherous waters to prove his spiritual power and vision of peace to skeptical audiences. This feat is not to be overlooked: it is a pivotal moment for the acceptance of his message of peace and the movement for forming the Haudenosaunee Confederacy.

The way rocks and stones figure into these stories could be dismissed as a nonessential feature of the narratives, yet they contain a silent power that is not to be ignored. Their presence holds a protective element, a source of wisdom, a static steadfastness that is always there for us humans. As Indigenous people, we respect their connection to deep historical and geological pasts. Stones carry knowledge, as do the stories. Our stories encourage us to also consider human nature, conflict, our challenges, and the responsibilities we carry to the earth and to one another as people.

Under a Smoky Haze: A Climate in Crisis and a Message

We may want to consider rethinking our relationship with land and our place as inhabitants of Turtle Island. We are similar to the fossils, an imprint in the great passage of time and space, yet as humans, we were gifted the unique abilities of consciousness, reason, and responsibility. The Dish with One Spoon refers to a wampum belt and treaty agreement called the Sewatohkwàt:shera:'t, which translates to "one spoon." The treaty's intent was to prevent warring nations' violent conflict over resources, hence the use of a dull spoon as a utensil, as opposed to a knife, which could be used to cause bloodshed.

The wampum belt itself is a long belt woven with white and purple wampum-shell beads and strings to make a visual design. In the center of the mostly white belt, a round purple shape represents a bowl. The common bowl, a frequent reference in treaty relations, represents the egalitarian sharing of hunting grounds and resources, and supports forging alliances of peace, solidarity, and friendship. The Dish with One Spoon concept instructs us to live in a sustainable way, to not take too much and to give consideration, leaving some for others in the future. This kind of forward-thinking ethos acknowledges empathy, generosity, and a practicality for keeping peaceful relations, fostering a mindset and decision-making that keeps both the balance of creation and the security of future generations in the foreground.

Are we still living in this mindset? Have we taken too much from the land? We are experiencing the highest-ever recorded temperatures worldwide, major storms and unprecedented flooding events, and inextinguishable summer wildfires. When will we take responsibility and act responsibly? In a time when we perpetually extract and regard the earth as a disposable commodity, how can we act to resist the systemic oppression rooted in the illness of settler colonialism?

I write these words as I sit in the house in midsummer, watching an ashy haze overcome my neighborhood with smoke and particulates from the Quebec wildfires. Photos shared across the internet show an unusual orange-hazed New York City skyline, reminiscent of the vast pollution of the 1970s. With air purifiers running throughout the rooms and smoke still infiltrating my living space, while taking inhaler puffs for my asthma, I sense the alarm of climate change on a new, very personal level, and I feel deeply saddened and anxious. I sense the grief of an earth changing. There is a new word for this kind of grief now, *solastalgia*: "the distress that is produced by environmental change impacting on people while they are directly connected to their home environment."[10]

When I begin to feel despair about our state of being on this earth, I often return to a Haudenosaunee Confederacy address made to the United Nations in 1977 that was published as the series of writings *Basic Call to Consciousness*. It was written primarily by Seneca scholar Sotsisówa John Mohawk on behalf of the collective

Haudenosaunee Confederacy. The address was a pivotal moment in the midst of a resurgence of Indigenous activism, an urgent calling to the world to consider the immense ecological damage done to the earth over the previous four hundred years. Its message urges people to unite as a collective to ensure the protection of the earth and human survival for future generations.

The *Basic Call to Consciousness* asserts that we can find our way back to a sense of balance through reclaimed spiritualism and traditional teachings. The foundation of Haudenosaunee spiritualism is a deep relationship rooted in love and respect for the homelands upon which we live. Sotsisówa John Mohawk acknowledges: "Our roots are deep in the lands where we live. We have a great love for our country, for our birthplace is there. The soil is rich from the bones of thousands of our generations. Each of us was created in those lands, and it is our duty to take great care of them, because from these lands will spring the future generations of the Onkwehón:we. We walk about with a great respect, for the Earth is a very sacred place."[11]

The message reminds us that what we are experiencing is an environmental crisis stemming from a significant difference in philosophies. Most Indigenous cultures have practiced an egalitarian relationship with the earth and parts of creation, whereas many Western cultures have adopted an exploitative ideology rooted in settler colonialism, one that regards the earth as a lifeless object. Other portions of the writings delve into the roots of these differences, exploring human history and the beginnings of colonization. We are reminded that at one point in history, every person on this earth had Indigenous ancestors, meaning people who lived on, of, and closely to the land for their survival. The origins of our dilemma, Mohawk argues, is in the fact that modern people have strayed so far from this land-based existence that they have forgotten their connection and relationship with the earth.

There is an urgent yet timeless quality to this address of which Mohawk is acutely aware: "It is a call to a basic consciousness that has ancient roots and ultramodern, even futuristic, manifestations."[12]

Nearly fifty years after this address, we still see ecological crises at play, and some that are nearing a state of emergency.[13] In 2009, eighteen American scientific associations presented a statement confirming that, on the basis of scientific and scholarly research, climate change is underway.[14] Ocean waters are steadily warming and changing ecosystems.[15] Global crises such as wildfires, flash floods, and slow drought events are more frequent.[16] The global food supply and food security is threatened.[17]

The Dish with One Spoon teaching implores us to remember that the earth's resources are finite and should be used cautiously and sustainably. This message implies that we must circle back to our original philosophies and teachings about the earth. Mohawk scholar Amber Meadow Adams points to additional troves of knowledge about ecological loss, recovery, and teachings within the Haudenosaunee Creation narrative. In examining the narrative more closely as a response to both human and ecological loss, she concludes, "The global human commitment to comprehend and grieve catastrophic ecological loss and then act to recover a functional ecology remains uncertain."[18]

We Are Future Elders

Several years ago, on a rainy July afternoon, I sat under a tent at the Great Law Recital at the Onondaga Nation community, listening to the teachings shared by elders from across all the Haudenosaunee territories. Mohawk elder Sakokweniónkwas Tom Porter spoke and he reminded us that we should always consider the measure of time of seven generations into the future in our thinking and decision-making. He said to think about the "coming faces," the future generations ahead that will one day emerge from the earth. We were urged to remember the responsibility we have for the children, grandchildren, and great-grandchildren yet to come.

Although I am not a mother, I have younger cousins to whom I am an auntie. I am fearful for their generation and generations to come and the world they will inherit. I feel saddened, powerless, and frustrated at the oppressive and exploitative systemic powers greater than myself. I can imagine that my parents and grandparents shared similar anxieties.

And still, I hold close the teachings shared with me by my parents, my grandmother, and my community. Gardening and cooking have been nourishing and healing ways for me to cultivate a relationship with the earth. When I was a youth, my grandmother showed me how to save seeds, when to plant, how to tackle gardening challenges, and how to preserve foods through water-bath canning. She taught me important self-sufficiency skills and tools for survival during hard times, yet also how to hold a love for the land and how it sustains us.

Thousands of years from now, what history and knowledge will the stones hold in their layers about us? What kind of individual and collective imprints are we leaving as humans in the greater geological and historical record? How can we carry good stewardship and good minds in our responsibility as human beings here and now? How do we restore our relationship with the earth and the lands and walk upon the rocks with care?

The landscape of the Niagara Escarpment has also been an especially meaningful way for me to connect with ancestral knowledge and access healing. Roaming the regional terrain of deep Carolinian forest, foggy meadows, steep earthen paths, and winding riverbanks was my medicine. I later came to appreciate my father's geology lessons and how they helped me better understand my place as a human within the world and all of its creations. To me, being Onkwehón:we, "a real person," a human, means holding a deep kinship and responsibility to the earth and understanding the ways the land provides all that we need to survive.

Every year in late winter, I return to my seed collection and begin planning my garden for the next growing season. Sorting through different strains of corn, beans, tomatoes, sunflowers, and tobacco, I select seeds to start. Watching them sprout and grow in little containers in the cold, dark basement brings me hope. I think about my ancestors who tended to the seeds and made it possible for us to continue to grow them today. Tuscarora people, my ancestors, migrated from North Carolina to western New York and carried seeds with them along that journey. They knew how to preserve our traditional foodways and plan for food security for future generations to survive. They knew that our seeds and foods represent hope. To save seeds means to harness responsibility and to invest in a future.

The riverside journeys with my father, gardening and preserving food with my grandmother, and the way the land shaped me have nourished me. The stones braced me on steep climbing hikes. The boulders quietly sat behind me while I meditated. The faded ridges of shelled creatures in stone layers were all tiny teachers to me. These are the moments that shaped the love and care I hold for Yonkhi'nihsténha Onhwéntsia. This translates to "she is mother to us, the earth" in the Mohawk language, with the root for *mother* coming from the concept "she lends her power."[19] Our Mother. The Earth.

What brings some solace and hope is the seeds and an ongoing connection to our traditional foods and their transformative capacities. As long as we continue to grow, learn, share, teach, and continue these traditions, these are ways we can retain our power. It's difficult, surreal, and often intimidating to accept, but eventually we may come to realize that we are the future elders in the making. We are future ancestors in a long lineage. By acknowledging this responsibility, we start to heal and create a better future.

notes

1. Deana Schwarz, "Geoscience Today: The Niagara Gorge Geotrail," *Beneath Your Feet: A Geoscience Blog*, January 31, 2023, https://geoscienceinfo.com/the-niagara-gorge-geotrail/.
2. "Geology and Niagara Falls," New York State Museum, https://www.nysm.nysed.gov/research-collections/geology/resources/niagara-falls.
3. "Natural Phenomenon: Niagara Falls," Niagara Falls National Heritage Area, https://www.discoverniagara.org/natural-phenomenon.

4. Stephen Vermette, *Weathering Change in WNY: WNY's Five Climate Zones*, Designing to Live Sustainably, December 2017, https://weather.buffalostate.edu/sites/weather.buffalostate.edu/files/uploads/photos/PDF/WNY5ClimateZones.pdf.

5. James Macauley, *The Natural, Statistical and Civil History of the State of New York*, vol. 1 (Albany, NY: Gould and Banks, 1829).

6. Edward G. Lengel, ed., *The Papers of George Washington*, Revolutionary War Series 20 (1779; Charlottesville, VA: University of Virginia Press, 2010), 716–719.

7. Arthur C. Parker, *Seneca Myths and Folktales* (Lincoln: University of Nebraska Press), 97–100.

8. Sotsisówa John Mohawk, *Myth of the Earth Grasper: Iroquois Creation Story* (Buffalo, NY: Mohawk Publications, 2005).

9. Mohawk, 26.

10. Glenn Albrecht, Gina-Maree Sartore, Linda Connor, Nick Higginbotham, Sonia Freeman, Brian Kelly, Helen Stain, Anne Tonna, and Georgia Pollard, "Solastalgia: The Distress Caused by Environmental Change," *Australas Psychiatry*, no. 15 (2007): S95–S98; Madeline Ostrander, "The Era of Climate Change Has Created a New Emotion," *The Atlantic*, July 23, 2022, https://www.theatlantic.com/science/archive/2022/07/climate-change-damage-displacement-solastalgia/670614/.

11. Akwesasne Notes, *Basic Call to Consciousness* (Summertown, TN: Native Voices, 2005), 86.

12. Akwesasne Notes, 83.

13. Julia Rosen, "The Science of Climate Change Explained: Facts, Evidence, and Proof," *New York Times*, April 19, 2021, https://www.nytimes.com/article/climate-change-global-warming-faq.html; Shannon Osaka, "Why Many Scientists Are Saying Climate Change Is an All-Out 'Emergency,'" *Washington Post*, October 30, 2023, https://www.washingtonpost.com/climate-environment/2023/10/30/climate-emergency-scientists-declaration/.

14. "Scientific Consensus: Earth's Climate Is Warming," National Aeronautics and Space Administration, https://climate.nasa.gov/scientific-consensus/.

15. "Vital Signs: Ocean Warming," National Aeronautics and Space Administration, https://climate.nasa.gov/vital-signs/ocean-warming/.

16. Xing Yuan, "A Global Transition to Flash Droughts under Climate Change," *Science* 380, no. 6641 (2023): https://doi.org/10.1126/science.abn6301.

17. Amanda Little, "Climate Change Is Likely to Devastate Our Food Supply. But There is Still Reason to be Hopeful," *Time*, August 28, 2019, https://time.com/5663621/climate-change-food-supply/.

18. Amber Meadow Adams, "Yotsi'tsishon and the Language of the Seed in the Haudenosaunee Story of Earth's Creation," *English Language Notes* 58, no. 1 (2020): https://doi.org/10.1215/00138282-8237454.

19. Sakokweniónkwas Tom Porter, *And Grandma Said… Iroquois Teachings as Passed Down through the Oral Tradition* (n.p.: Xlibris, 2008).

The Hill Has Something to Say

Rita Dove

but isn't talking.
Instead the valley groans as the wind,
amphoric,
hoots its one bad note.
Halfway up, we stop to peek
through smudged pine: this is Europe
and its green terraces.

*

and takes its time.
What's left
to climb's inside us,
earth rising, stupefied.

*

: it's not all in the books
(but maps don't lie).
The hill has a right
to stand here, one knob
in the coiled spine of a peasant
who, forgetting to flee, simply
lay down forever.

*

bootstrap and spur
harrow and pitchfork
a bugle a sandal
clay head of a pipe

*

(For all we know
the wind's inside us, pacing
our lungs. For all we know
it's spring and the ground
moistens as raped maids break
to blossom. What's invisible
sings, and we bear witness.)

*

if we would listen! Underfoot
slow weight, Scavenger Time,
and the little old woman
who lives there still.

Rescuing Darkness

Jane Slade

I have come to know the sound of mourning doves as they land and take off. It reminds me of a screen door in the summer, something between a squeak and a whistle that becomes a familiar sound of coming and going. This sound is important if you want to know the mourning dove because you can geolocate where they are coming from and where they are going, and especially when they are coming home to their nest.

This is how I found the nest that is just beneath my window, which allowed me to witness a mother and father bring many pairs of fledglings into this world, sometimes going through the process of egg laying to first flight six times in a spring. They taught me so much of what it is to be in a bird family: the body language and communication, the timing of their lives, the rhythm of how a day is spent and how a night is spent, the gravity of what it takes to survive, the expectations of a partner, a parent, a child, and a sibling, and namely, that there is no shortage of love and that animals, too, feel deep love and protection for their kin.

I was lucky enough to see a pair of fledglings make their first practice jumps from the nest to a nearby electric wire as the parents demonstrated it was possible. This process took a day or two to convince each of the baby birds, one a little more frightened than the other. On the evening of their final jump to the wire and their first night in the tree, the baby birds had yet to commit to the leap of faith into the branches above. Both parents sat on the wire, patiently coaxing.

Figure 1. The baby mourning doves are just to the left of their parents on the wire. The nest is behind them within the leaves. They have made the jump to the wire and are about to make the leap from the wire into the nearby tree.

Suddenly, a squirrel tried to commute by using the wire, and the presence of such a large animal compared to the baby birds was about to disrupt everything. On instinct, the mother immediately made her body seem bigger by creating a very uncomfortable geometric shape with her wings, so much so that the squirrel turned right around and avoided the confrontation altogether. Soon enough, the baby mourning doves, or squabs as they are called, each took the leap to fly up into the tree.

Watching a baby bird learn to fly was different from what I had thought. It was not a sink-or-swim moment; there was a buildup, a unified family focus, support, and love. The squabs did not walk off the gangplank and hope for the best. They practiced, flapped their wings to stretch, worked up their courage, focused. They turned outward and considered their leap for hours. The parents were there watching their children, guiding, coaxing, ensuring, and filling in all the gaps.

I was able to track the babies the next day, which was also not full of perfect flight. Their parents stayed near to guide and help

Figure 2. The mother dove getting big to protect the squabs' first flights from interference by the squirrel.

them take flight within this known terrain. One ended up on the ground with one of the parents helping. Watching this taught me that it is not an accident that birds fly; there is great teaching involved, and the parents ensure that proper flight skills are learned.

Obsolete Instincts

As I grew to love the mourning doves, my heart ached when each fledgling would find their way out of the nest. Not only because I had become attached to the presence of this little bird family but also, because of my research on light pollution, I knew how deadly that danger loomed over them and how the world at night no longer resembled the instincts encoded in their genes—to migrate by the map of the stars to chosen lands and to sleep, eat, prey, and procreate by the light of the sun and the moon.

I remember being a kid and holding my hand over a small flashlight, my hand aglow. Light at night penetrates tiny bird bodies just the same, sending discordant hormonal messages and misaligning nature's cues. We are see-through to the light, bird bodies, human bodies—all living creatures. Birds suffer immensely because of light pollution, which turns night into day, and they are the canaries in the coal mine telling us that we are doing something wrong.

Each year, incalculable numbers of birds fatally crash into our glass architecture. Not only have we blocked a view of the stars through skyglow, but the faux constellations of light-polluted cities distract and disorient migrating birds at night, leading to deadly collisions that can sometimes kill hundreds of flocking birds nearly instantaneously. All across the world, birds are in serious peril from climate change, with 49 percent of species in decline.[1] In North America, we have lost nearly three billion birds since 1970.[2] The brightening of the night is yet another way we have distorted life on Earth for migrating birds. Humans threaten even more disturbance with perversely unnatural plans like massive satellite billboards floating through the night sky. Yet the sky is very much their home. In fact, some birds spend more of their lifetime flying through the sky than they do on the earth.[3] They make these long and treacherous journeys over oceans and land in order to complete the cycles of their lives.

The sky was the first screen, providing vital information about the time of day, the seasons, the weather, and migratory trajectories. Birds, among many other migratory species, including whales and dung beetles, utilize the map of the stars to find their way on the planet.[4] Light at night is a human manipulation of data, falsifying and redacting once-perennial maps and flippantly distorting the rhythms of life on Earth that were nudged into balance over eons. Starlight is a matter of survival. Darkness is an ecological necessity.

In fact, night is older than Earth itself. Over four and a half billion years ago, Earth formed from spinning matter and gas that slowly condensed into a turning planet. We speak so much of global warming but say almost nothing of light pollution. Light and temperature are two fundamental environmental factors that have tuned the evolution of all living things. We are rapidly changing both.

Modern lighting practices have made the day eternal. Currently, the world is experiencing exponential increases in light pollution each year due to the proliferation of LED lighting, which requires very little electricity in comparison to earlier light sources.[5] When we lost the economic incentive of turning lights off to save money, we stopped turning lights out altogether and then added even more light at night into the environment. Moreover, the intensity of LED fixtures can create great harm across vast distances, from the general disruption of darkness to skyglow and glare; it is too easy to pollute with just one light. If you look at light-pollution maps, darkness has been virtually erased from large swaths of the globe, obstructing a view of the night sky for over 80 percent of humanity.[6] We are extinguishing the night.

The impacts on wildlife are haunting and heartbreaking. There is no species that is not affected. Nocturnal pollinators die from distraction, like moths to a flame; bats become fixated upon eating the distracted insects at streetlights; age-old predator-prey relationships shift; disoriented baby sea turtles never make it from the beach where they hatch to the water for safety; zooplankton stop vertically navigating to the surface of waters because the message of darkness never comes; birds reproduce prematurely; trees develop leaves prematurely; plants produce fewer fruits; and many more grave impacts occur that set the beautiful and complex inter-species interactions of Earth's diverse ecosystems askew.[7] As these changes cascade through our ecosystems, the harm becomes incalculable.

The impacts of light pollution on wildlife are not yet well known to the public, yet it is the one form of climate change we could solve with the flick of a switch. Awareness is the biggest barrier. Sadly, the well-being of wildlife has never been enough to turn the tide of human industry, in spite of the fact that our own survival is woven together. Perhaps to better galvanize dark-sky advocacy we should not just be asking, What are the impacts upon wildlife? but also, What are the impacts of light pollution on humans, and what are we missing when we stop experiencing natural darkness?

An Epidemic of Disconnection

The presence of light at night has changed human experience, too. Our attempts to control the night with light have blinded our connection to the universe and radically changed our own relationship to light and darkness here on the earth. Instead of the red light of fire, we shine blue light, the color that shines at the zenith of the day, directly into our eyes before bed. The ubiquity of screens has robbed humans of the healing benefits of natural darkness, shining light into the windows of our brains when we should be resting and recovering. In the past, we awoke to the ruddy light of sunrise, hunted and gathered under the blue light of the sky and the green light of trees, cooked our food by the red light of fire, and when the fire dimmed, the stars came out under a navy and purple sky. These experiences of natural darkness and the rhythm of colored light have become nearly extinct within our lived experience.

Instead, we spend most of our waking experience in bright light. Light-driven information can find us at any time, and we are suffering from digital burnout. When we never let our days finish, we blockade moments of reflection, and our hearts and minds become cluttered. This disconnection from darkness brings immense stress and furthers disconnection from our intuition. We are also missing out on modernity's promise of more ease and time with our loved ones. The natural daylight cycle is the Rosetta stone to find our way back from light pollution.

If there were one message I would leave this world with, it would be this: the natural daylight cycle at all points is a gateway to the present moment and a connection to all living things that have ever lived on Earth.

Light at night is a virtual experience, a disconnection from the experience of the natural daylight cycle. There is an epidemic of disconnection happening all around us, and we have forgotten the parts of ourselves that are still wild, what it is to be a living thing on a turning planet. An aversion to darkness is a modern affliction and perhaps misguides our expectations that all darkness, both physical and metaphorical, can be avoided. This aversion also undervalues a critical period of rest, reflection, and self-exploration. The darkness of the night and the darkness of our souls are not to be avoided but explored.

Natural darkness is nightly medicine for our physical, mental, emotional, and existential health. As daylight recedes, our visual systems become cloaked in darkness, turning off this massive form of stimulation to our brains and shifting the hormonal landscape within our bodies. Natural darkness provides natural sensory deprivation, shifting human awareness from a mind-centered experience powered by vision to a body-centered experience. If light turns you into a receiver, then darkness turns you into a transmitter. When the lights go out, internal reflection becomes possible—to find your place in the context of both daily life and the universe.

Humans rule the day, but we will never rule the night. We may be ruining the night and risking ourselves and other living things in the short term, but natural darkness cannot be overpowered by humanity. Moreover, exploration of the dark is a fundamental aspect of life on Earth. When the stars come out, all living things undergo a realization of the infinity above our bodies, confronting our very existence. Seeing through the eyes of starlight bestows infinite points of view. This direct connection to the unknown and the unknowable has been a perennial source of awe and inspiration for humans. When we consider all that we do not know, it becomes inspiration for problem-solving. For many, starlight has ignited epiphanies, transforming inhibition into inspiration.

Still Wild

I did not always know the intimate sounds of birds. It was taught to me by a long lineage of birds, many who are now ancestors, whose many life spans occur within my one lifetime. I picked up the nuances of their tongues as I was sleeping, their song drifting into my window from the treetops surrounding my urban apartment. When I first moved into my home, I found the birdsong so unusually loud for a city that I bought a white-noise machine. Now the birdsong is the white noise. I did not know any of their calls, but now I know the robin's sweet song, the cardinal's pip—which feels like a secret message that they are sending to one another—and the mourning dove's woody and rolling incantations. Even as I write this very sentence, four turkeys are foraging below my window in the middle of Cambridge.

The story the mourning doves told me was a gift of the highest order, given with no expectation of return, with no expectation that I would even notice. Wildlife is all around us, breathing lessons of connection and understanding, and yet we hardly realize it. We have forgotten to witness these parts of ourselves that are still wild too. Yet what they taught me was a path back to these untamed places—to the parts of me that live by the light of the sun and moon, to the parts of me that fight for survival if only to love, to these same parts of all living things. While birds around the world are telling us that we must rescue darkness through heartbreaking loss of life, in my own life, I awaken to the cooing of mourning doves telling the wordless story of their lives, of fellow earthlings daring not just to live but to thrive in this wild world. There are natural guides all around us if we are willing to listen.

The experience of night is rich with the most precious moments of our lives, from breaking bread with our loved ones to tucking our children into bed. To revel in the darkness of the night is to be fully alive—to pursue not just happiness but also what it means to fully live out a human life. It's not that we are unable to see in darkness; it is that darkness is for another way of seeing altogether. When the lights go out, our forms soften, and we share some of the most vulnerable and meaningful parts of ourselves with one another.

When we can no longer see our own bodies, natural darkness makes it possible to return to the place of our ineffable existence.

There is a place within that is only yours, where the light always shines. I know this because I have come to this place through two different doorways. Through the darkness that comes with living in a mortal body that loves, is left, and leaves, and through the darkness that occurs each night on this planet and the experience this creates by blurring the boundary of my body altogether. In darkness, you can erase the self and begin again when the light comes. Each night, this journey is possible.

notes

1. BirdLife International, *State of the World's Birds 2022: Insights and Solutions for the Biodiversity Crisis* (Cambridge, UK: BirdLife International, 2022).
2. Kenneth V. Rosenberg et al., "Decline of the North American Avifauna," *Science* 366, no. 6461 (2019): 120–24, https://doi.org/10.1126/science.aaw1313.
3. Felix Liechti et al., "First Evidence of a 200-Day Non-Stop Flight in a Bird," *Nature Communications* 4, no. 1 (2013): 2554, https://doi.org/10.1038/ncomms3554.
4. On humpback whales, see Travis Horton et al., "Straight as an Arrow: Humpback Whales Swim Constant Course Tracks during Long-Distance Migration," *Biology Letters* 7, no. 5 (2011): 674–79. https://doi.org/10.1098/rsbl.2011.0279. On dung beetles, see James J. Foster et al., "Light Pollution Forces a Change in Dung Beetle Orientation Behavior," *Current Biology* 31, no. 17 (2021): 3935–42, https://doi.org/10.1016/j.cub.2021.06.038.
5. Christopher C. M. Kyba et al., "Citizen Scientists Report Global Rapid Reductions in the Visibility of Stars from 2011 to 2022," *Science* 379, no. 6629 (2023): 265–68, https://doi.org/10.1126/science.abq7781.
6. Fabio Falchi et al., "The New World Atlas of Artificial Night Sky Brightness," *Science Advances* 2, no. 6 (2016), https://doi.org/10.1126/sciadv.1600377.
7. See Eva Knop et al., "Artificial Light at Night as a New Threat to Pollination," *Nature* 548, no. 7666 (2017): 206–9, https://doi.org/10.1038/nature23288; Jens Rydell, "Bats and Their Insect Prey at Streetlights," in *Ecological Consequences of Artificial Night Lighting*, ed. Catherine Rich and Travis Longcore (Washington, DC: Island Press, 2006), 43–60; Marianne Moore et al., "Urban Light Pollution Alters the Diel Vertical Migration of Daphnia," *Verhandlungen Internationale Vereinigung für Theoretische und Angewandte Limnologie* 27 (2000): 779–82, https://doi.org/10.1002/9780470694961.ch1; Davide M. Dominoni et al., "Artificial Light at Night Advances Avian Reproductive Physiology," *Proceedings of the Royal Society B: Biological Sciences* 280, no. 1756 (2013): art. 20123017, https://doi.org/10.1098/rspb.2012.3017; Monika Czaja and Anna Kołton, "How Light Pollution Can Affect Spring Development of Urban Trees and Shrubs," *Urban Forestry & Urban Greening* 77 (2022): art. 127753, https://doi.org/10.1016/j.ufug.2022.12775.

Meditations on Eels and Earth

Kristi Leora Gansworth

An eel is a gentle and beautiful teacher. In some company, the words *eel* and *beautiful* do not go well together. In other settings, eels are understood as deities, worthy of reverence and honor, whose presence invokes a supernatural encounter. In my time as a researcher and avid fan of eels, I have witnessed both the awestruck eel enthusiast alongside the disgusted loather of eels, and a spectrum of opinions in between. I have learned, from a range of people, about attitudes and beliefs toward specific animals, revealing insight into broader relationships with the planet and the many creatures who make Earth a home of pluralism and difference.

Scientific research suggests that eels have been living on Earth for perhaps seventy million years, longer than many species alive today. They have survived times of mass extinction events, transitions where Earth conditions become livable for some, unlivable for others. The climate, chemistry, and composition of Earth's layers and systems change over time. Countless species have disappeared, shifted, or receded, and some persist. Earth's evolving stories are deeply embedded in stored planetary memory: fossils, rock, clay, soil.

Some methods for understanding the life history of eels come from that physical memory: information and speculation are gleaned from eel fossil imprints and bones. Ideas and theories appear in a corpus of written material, a centuries-long effort to understand how eel migration works. When it comes to a definitive narrative of eel reproductive cycles, however, that research offers few solid answers. Eels have complex life cycles and migration journeys that leave unanswered questions, although there are some assumed certainties. One collective and especially striking observation: in recent years, all members of the genus *Anguilla* have experienced a worldwide decline.

The eel to whom I refer, *Anguilla rostrata*, is a relative of eighteen other members in the genus. These are fish whose lives are characterized by metamorphosis. All Anguillid species are catadromous, meaning they begin life in the ocean and travel into fresh and brackish waters, often found in estuaries but also in freshwater streams, lakes, creeks, ponds, and other areas.

Along their migration journey, which could be two thousand miles or more, eels adapt to the changing conditions as they travel, often at night. They are guided by an internal compass connected to Earth's magnetic fields, and they also navigate and swim in different ways depending on where they are, following cycles of the moon and stars. Observing eel migration enables one to keep close track of the changes in seasonality, including conditions in spring, when new eels migrate into watersheds, and in summer toward fall, when adult eels leave the various watersheds and return to the ocean.

Maturing eels can do amazing things: they can survive out of water by moving across land, they can form masses with other eels to collectively move together in and out of water, and they can climb waterfalls. In the cold seasons, eels burrow in mud and go dormant, their heartbeats slowing to a dull, deep pulse. Like seasons of the Earth, winter calls their bodies into a state of rest and stillness. Their movements and behavior, at times, correspond with seasons in fascinating ways, attuned to the rhythms of Earth.

Endangerment

There are a good number of places along the eastern seaboard where eels continue to migrate and return in large numbers, even if unevenly. There is debate about whether there should be commercial eel fisheries and who should run them. Eel habitat, which was once managed by Indigenous societies through legal systems and kinship structures, has been highly developed and privatized. The result is a loss of range and space, which once enabled their constant, year-round abundance. Old stories about eels from colonial records and other writings suggest that eels could be fished by the thousands in a single night.

Eel declines in North America are a result of disruptions originating in modern human societies. Altering water and land on a large scale has led to eel depletion, endangerment, and even extirpation in some watersheds. Examples of these alterations include the installation of water control systems: irrigation canals, culverts, dams, and related European agricultural infrastructure. Eel migration is also affected by the drainage of wetlands and other profitable schemes that create obstructions and block migratory fish and other animals from traveling their ancient pathways.

The turbines of hydroelectric dams are especially daunting for eels. Sharp, spinning blades that create velocity in the water to power hydro facilities often create piles of shredded fish who never make it past the structures.

Concerns have been raised by multiple parties since the 1980s about alarming rates of eel disappearance in North America and beyond. Indigenous elders familiar with eels have indicated that they see major changes to the viability of fish migration, and they express great concern about the wellness of animals who they have known and depended on for generations.

The reduced presence of eels around the world is also connected to the pollution of water; the human trade and movement of opportunistic, predatory aquatic species and parasites; sea temperature change; and other global shifts among Earth's constituents.

In short, there is no singular factor to blame for the decline of eel populations. Efforts have been made to ameliorate the adverse impact of hydrological changes and improve human systems through technologies such as fish passage structures, which have some benefits but are not a cure-all. It seems that eels are specifically impacted by human societies in profound ways that have taken decades to reach a zenith. Whether they should be listed as endangered or legally protected in other ways is a matter of debate across jurisdictions: the United States and Canada (and their states and provinces) have different practices and perspectives.

About Eel Life

It is important to get a sense of how expansive, intricate, and fascinating eel migration and metamorphosis are. *A. rostrata* begin their lives in the deep Sargasso Sea and travel in a northwestern direction, going through distinctive life stages. Their historical range stretches as far north as Greenland, and they were known to travel throughout and past the Mississippi River watershed in prior generations. Eels are masters of navigation, moving through the inner contours and caverns of Earth with a skillfulness that is based on millions of years of familiarity and repetition.

In their first stage, eel larvae look like tiny eggs, floating north along the Gulf Stream with a large chain of migratory organisms, moving north and trailing off into watersheds along the North American seaboard. This influx of new larvae arrives each spring, and the tiny eggs grow into a leaflike shape, the leptocephalus stage. Next comes the elver stage, which leads to a yellow stage and finally to an adult, silver stage.

Each life stage can last anywhere from one to five years, and each stage involves a series of dramatically different shapes and sizes, which can make eels difficult to identify for those not familiar with their changing form. After spending several years in estuaries or freshwater, silver-stage eels migrate back toward their site of origin in the Sargasso Sea, presumably to spawn and die.

Eels are intimate guides to Earth's watersheds: they touch the deep oceans, navigate shallow shorelines, and appear in small water bodies. They come from massive and nebulous marine zones. Like every eel fanatic, I can tell you many more facts and details about eels, and yet there are stories, more personal in nature, that I'd like to share as well, because facts are only one way to know about eels. There are other relations that matter.

Embodied Connections

I refer to eels as my teachers because, as I understand things through an Anishinabe lens, animals are kind and devoted teachers who have their own forms of personhood. Each has an idiosyncratic intelligence and purpose with particular contributions that they make in the united web of life. My belief is that animals also have a spirit, as do all creatures, and that spirit is a sacred expression that must be respected.

Specific stories and orientations of Indigenous peoples and their governance structures differ. Connections to animals, place, and prior generations remain living memories from which to draw a wider understanding of possibilities. I am Anishinabekwe and belong to a People whose lives are sustained by the diversity of seasons, the gifts, and the generosity of Earth.

Pimisi is a word that translates as "eel" in the language of Omamiwinini Anishinabeg, which means "the people up and down the river" according to my late grandmother. Others call us Algonquins, a People who have lived along the Ottawa River since time immemorial, my maternal relatives and ancestors. Eels were the foundation of fishing cultures, ceremonial cycles, and trade relationships in that territory. Today, there is a light-rail station in Ottawa named Pimisi, named to honor the eel presence in the watershed. Nearby, a sculpture sits in the water, a tribute to the Anishinabeg and their knowledge of and respect for the life cycle of the migratory fish (fig. 1).

The sculpture, a twenty-five-foot-high eel, is formed in stainless steel, drawing attention to the ways the city and its infrastructures cut through eel habitat. The sculpture demonstrates the fragility of all that exists in the built environment alongside overlapping ecological webs of the Ottawa River. The sculpture is a reminder that eels once composed a significant percentage of the river's biomass, and their presence in the river persists despite the massive habitat alterations and desecration of Indigenous territories that is so commonplace today.

Figure 1. Nadia Myre, "Untitled (Pimisi/Eel)," OLRT Pimisi Station, City of Ottawa, 2018. Photo by the author.

There are numerous sites in the capital city that contain the memories of Anishinabeg predecessors and ancient relatives, sites that have also changed from their prior forms. It is my desire to honor the ancestors I come from whose everyday lives were closely intertwined with water, Earth, and eels, living a different way of life than what I see and experience today. In generations past, *pimisi* and other migratory fish, birds, and animals were the ones who taught us how to survive and live in relationality with our surroundings.

Eels sustained generations of Anishinabeg and our neighbors and relatives, including the human and nonhuman newcomers that now make these lands their home. It is difficult to experience the cognitive dissonance that comes from witnessing abuse and alteration of the Ottawa River and all the nonhuman residents who live there. A new default has been established by building over the history of the days when eels filled the watershed.

A complex cultural history in Ottawa is sometimes difficult to apprehend amid the changes resulting from urban development, including the removal and replacement of Indigenous peoples and the disregard of ecological integrity that my ancestors and relations contributed to for several generations. To trace my ancient, ancestral memory, I keep journals and running meditations to capture what different sites mean to me and how they intersect with the layers of my own knowledge and being. The violation of sacred areas and grave sites is a running and persistent issue. I share here an excerpt from one of my journals, written at a site that has been earmarked for development despite the protests and warnings of Omamiwinini Elders.

A Meditation: The Sacred Site Remains

Eel bones and human bones rested here together. This is not the haphazard scattering of the dead affiliated with a massacre, trauma, and placelessness: it is deliberate. It is the marking of a seasonal gathering, offerings of thanks, demonstration of skill. These are the rungs in a barely perceptible web, one which is cut through by human hands who pull from the Earth to create their surroundings, sometimes in purposeful ways, sometimes in destructive ways.

The land remembers: people sing, feast, visit, a tree wears crowns of flowers and gifts placed by descendants who pore over the sorrow and determination they carry, to remember, to re-learn, to reciprocate what has always been on offer: a life informed by movement of celestial bodies including the moon, the stars, and other inter-galactic relatives.

Fishing camps and traplines, medicine gathering sites, ceremonial sites: these places, stories, and traditions are embedded in the landscapes. Our history and our future originate here. There is devastation and demolition, exploded rocks, disturbed caves, exposed burial sites, threats of nuclear waste, and plundered graves. There is much offering to make and much to stand for if life is to continue.

Indigenous Laws and Intergenerational Obligations

In multiple Indigenous cosmological orientations and societies, care of the living and of the dead is important to maintaining balance between spiritual and physical realms. Clans form important political and spiritual identities related to these forms of balance. On the other side of my family, my Onondaga father belongs to the eel clan, like his mother—my grandmother—and all his maternal grandmothers going back several generations. Clans are foundational to Haudenosaunee social organization, law, and kinship. Haudenosaunee people alive today can trace their ancestors through lineage all the way to the time when the Great Law of Peace was brought to their people, at least four hundred years ago but likely much longer.

Like our Haudenosaunee kin, Anishinabeg honor and carry relationships with clan animals that create intergenerational cycles of kinship, reciprocity, and accountability. These are also the values that inform our legal traditions, which are embodied, lived, and experienced through connections with the natural world and with ancestral continuity.

For Anishinabeg, being affiliated with a clan is a lifelong responsibility. Animals to whom Anishinabeg are most closely related are our clan relatives: they create the structure of a person's life path and relationality. One's animal helpers from families on both sides are lifelong helpers and guides. Clan relations (or *doodemag*) are inherited, or adoptive, not chosen or a matter of personal preference.

In my current work, I make efforts to understand living conditions for the animals who belong to my ancestral lineages. I believe in that obligation. I believe that their time on Earth has preceded ours by countless generations and their lives can teach us meaningfully if we attune to the realities they experience.

Each animal carries different meanings: what we learn from wolves is very different from what we learn from eels, for example, yet both have a place and meaning in the specific environments where their impact is most closely felt. All animals carry medicine, and that medicine is to be respected and understood as a matter of law, reciprocity, and humility, which are central values in Anishinabe law as well as the related but distinctive legal traditions of our neighbors, allies, and relations.

The Offerings of Eels

One of the most fascinating aspects of an eel's journey takes place during the silver stage, when adults begin a long journey back toward the Sargasso Sea. As adults, their life cycle ends when they spawn, and it is believed that females release eggs in the wide-open ocean. There is a great deal of speculation about exactly how this happens—something that occupies the minds of many people.

For me, the more significant aspect of the silver stage is that migrating eels are in a fasting state. As they travel for sometimes hundreds of miles overnight, they stop eating, and scientists have found that as the eels migrate their bodies distribute nutrients to the food chain, nutrients stored in their guts, valuable and necessary nutrients that are sometimes found only in specific areas such as salt marshes.[1]

This means that as eels travel to the site of their assumed origin, their bodies return and transport viable material that supports oceanic life in meaningful ways. This is one of many instances in which eel migration demonstrates powerful forms of connectivity and contribution to larger spheres and cycles.

Eels have survived and borne witness to Earth's geologic eras, including times of fire, ice, tectonic shifts, continental drifts, and climactic changes. We know that adult eel bodies go into a state of fasting as they exit an estuary and return to the ocean, and we can consider that transformation through the lens of Indigenous land-based healing practices as well.

Fasting is an important way of knowing in ancient Indigenous practices; fasting is a time when individuals and communities voluntarily forgo the comforts of everyday life to pursue wisdom, purification, healing, knowledge, improved health, and other connections. Fasting has been a reliable method of wellness for Anishinabeg for centuries. It is also a way to cope with the "diseases of civilization" that "disproportionately affect Indigenous Peoples" in modern times.[2]

Learning from eels about the value of fasting changes the questions one asks: What does it mean to offer elements of oneself, to give of oneself for the benefit of others? What level of consciousness is activated? Fasting is a profound and intentional practice that enables close relationship with the rhythms of Earth, including the cycles of the Sun, the Moon, and the movement of plants, animals, birds, and trees—the holistic web of life.

Growing into adults, eels function as prey for larger animals such as herons, eagles, hawks, and otters. They play an important role as benthic and carrion feeders in waters, meaning that they help to keep waterways clear of debris. Eels develop connections with other organisms such as freshwater mussels, a symbiotic relationship that enables water filtration, which is important for maintaining clean water in freshwater habitats. These are only a few examples of the many amazing relationships that eels are part of. One's life, properly attuned to Earth, can be a gift to many and to the larger whole, to the cycles of continuity represented in Earth's seasons and other cycles.

Closing Meditation and Reflection

I have been formally studying eels for over ten years in different settings and have been a relative to eels since before I was born. I have listened to the words of artists, conservationists, Elders, youth, scientists, lawyers, venture capitalists—all with a range of perspectives. Some suggest that human behavior needs to change immediately and that the structures that enable modern life are to blame for eel declines. Some who are very close to the lands and waters say that the best use of our attention is to witness and observe the seasonal behaviors of eels to learn about how climate change and other factors affect their migration. Others suggest esoteric and spiritual approaches that seek holistic balance between humans and all our relations.

I have also listened to disgusted residents, newcomers to these lands, who despise eels and see no value in their lives. These are people who often know nothing about eel migration and make aesthetic or other types of judgments that have little basis in ecological knowledge and how eels literally carry the stories of Earth in their memories and their behaviors—their hearts, eyes, and guts.

The wish and devotion I continually make is for people of all ancestral backgrounds and relationships to think about humans as a relatively young species. I ask what it means to learn from those who are our Earth elders—animals like eels who tend places that humans do not control, can never fully know, and are obligated to respect.

Discussions about eels often raise ambiguities about ways to honor their historical and present importance, and my contribution is no exception. I close here with a short meditation that occurs to me each spring when eels and their relatives rush in from the Gulf Stream, symbolizing new beginnings and the start of another annual cycle.

Meditation: Leptocephalus Means Leaf

In the eye of my mind, beneath closed eyelids, the screen of my imagination shows a great tree on the ocean floor: My exhalation travels beyond ocean depths to fill the sky with the boundless respect that I have for this ocean, for this sky, for this life: respect for its mysteries, for its faithful and persistent tending of eels, squids, jellyfish, all the strange creatures that are branded alien, ugly, and unwanted.

A great tree lives in the bottom of the sea. Each baby eel that moves across ocean currents is a leaf from that tree, an extension from that tree of life. I say a tree because a tree is rooted, a tree builds other worlds, a tree can be formed by knots of fiber, of wood, of root, of hair. Through its decomposition, a leaf can give life to soil.

In the same way, an eel grows: first an egg, from which life emerges, then into a leaf and pieces of the body fall away into the streams of water. The animal grows and changes. A decaying leaf gives life to its own next stage.

notes

1. Alyson Eberhardt, "Rethinking the Freshwater Eel: Salt Marsh Trophic Support of the American Eel, *Anguilla rostrata*," *Estuaries and Coasts* 38 (2015): 1251–1261.
2. Michael Yellowbird, "Decolonizing Fasting to Improve Indigenous Wellness," *Cultural Survival Quarterly* 44, no. 2 (2020): 18–19.

Keeping Warm on Glacial Earth

Emma Gilheany

In a glaciated landscape, keeping warm is not only a mode of survival—as someone from warmer climes might think—but also a way of attuning to the relations between people, place, and Earth. This essay is about the work and joy of keeping warm in a specific, subarctic place: Hopedale, Nunatsiavut, a six-hundred-person community in the Inuit self-governing region of Labrador. Nunatsiavut—which means "Our Beautiful Land" in Inuttitut—is south of Nunavut and is bounded by the frigid Labrador Sea to the east. I first came to Hopedale in 2017 as a student of archaeology, staying for two months, and I have returned as much as possible since then. It is a journey that involves flights and a two-day ferry up the craggy coastline. Throughout my time in Hopedale, one of the many things I have learned is how a specificity of place—of relation to *Earth*—is borne of the deep geological histories that carved abyssal fjords and carried strata of dirt through millennia of glaciation, which is the process of the formation, existence, or movement of glaciers over the surface of the earth. Keeping warm in a place of intense glaciation is an ongoing practice that involves movement, place-based knowledge, labor, social visits, nourishment, and joy. I have come to call this phenomenon "social warmth" to encompass these ongoing practices. I share here a series of short vignettes, each grounded by conversations with different Hopedalimiut knowledge keepers, in order to describe what social warmth looks like for readers who are not accustomed to the circumpolar north.

Josephine and Her Woodstove

Past glacial movements continue to reverberate and shape the language and labor of every day. In Nunatsiavut, being "at the stove" means tending to the fire, being active in the creation of warming the home via a woodstove. Any time you visit someone, as you sip tea—customarily offered to guests—your host will be at the stove. I first noticed the importance of the woodstove when visiting an elder named Josephine, a small, bright-eyed woman whose entire family would become very dear to me and whose home would become the place I stayed when I was in town. On my first visit, I noticed that Josephine would intermittently shuffle from her upholstered floral armchair in the living room to the mudroom, grabbing splits of wood from an orderly stack made by her son, who arranged them neatly in a plywood box he had constructed and painted. Heeding a call inaudible to me, she would carry the splits to the woodstove, open the heavy cast-iron door decorated with an image of burning logs in bronze, and toss in birch or spruce, never pausing our conversation. She would poke at the fire sometimes, sparks visible, the wood crackling away all evening. The woodstove was something to tend to, something to care for, to stoke our own conversations, an ambient warmth that echoed that of Josephine herself. Generating such warmth is a way that people connect to each other that is bound to the capacities and the demands of Earth.

The woodstove is an object for thinking about social warmth and how warmth is forged by the environmental specificity of Hopedale. I am an archaeologist whose narrative methodology is rooted in the archaeological, in the careful removal of layers of the earth. The thin soil in Hopedale makes archaeological excavation particularly difficult. The windswept terrain forms a matrix of spongy lichens, mossy bogs, bare gneiss, and seasonal melt ponds often covered in ice and snow. In traditional archaeological excavation, the archaeologist documents not only the artifacts that emerge from the earth but also the earth itself, a process known as recording context. Archaeological context refers to the physical and environmental setting of an artifact—its location, its association with other objects and structures, the cultural and natural processes visible in its deposition.

I use the woodstove as an artifact, still in circulation, to think about earthen specificity and relationship to place. In this essay, I record the woodstove's context as an archaeologist does with a trowel, pencil, and paper: its provenience and its relationships with other artifacts, ecofacts, structures, systems, and people. This narrative is spun outward from the woodstove, a prism that refracts the experiences that have been shared with me, organized by conversations I have had with Hopedalimiut knowledge keepers. Archaeology is and can be a creative practice: for me, it is a sensibility, a methodology to crack open and document where bodies and the world meet, a material window into the constellations of sociality in the circumpolar north. Further, this type of artifact-narrative depends deeply upon immersion, upon oral history, upon relationships. The thin soil of Hopedale that frames my methodological conceit is part of what makes the woodstove an extraordinary artifact. It is an object that reveals place-specific labor, care, and long environmental histories that demonstrate the earthen nature of keeping warm as a dimension of relationality.

John, the Weather, and the Geological

Josephine's son John—known by some as the "weatherman of Hopedale"—has told me that glaciers were especially hard on this place compared to other communities on the coast. He has a remarkable appreciation and knowledge not only of long environmental histories in Nunatsiavut but also of the shifting weather patterns during his lifetime. He has kept a meticulous daily record of the climatic conditions in Hopedale for over three decades, housed in half a dozen notebooks. Each notebook is filled with neat, hand-lined columns that document the weather (separate entries for temperature, wind direction, precipitation, and related notes, such as the end of sea-ice breakup), beginning with his first entry in March 1992. John and I have discussed several times over the years that the treelessness of Hopedale is partially due to the particularly thin soil surrounding the community. There is not much for the roots to grab hold to because glaciers took the sediment with them on their ancient journey to the Labrador Sea.

Glacial erosion not only scrapes strata of earth to be delivered to the oceans, however. All that's left atop hard gneiss—a type of metamorphic rock—are the vestiges of pulverized rocks, a hard flour for tall flora to thrive in. Treelessness also means there is little buffer from frigid air currents that originate off the Arctic Ocean. This makes Hopedale rockier, colder, and windier than its neighboring towns—even more northerly ones. The geology and weather are in close dialogue here, stitching together a place of blustery storms, thick fog, and sublime vistas for the theater of the northern lights.

Hopedale is made up of boulder-strewn beaches, rocky cliffs, and rolling hills, undulations of three-billion-year-old gneiss carved by glacial erosion during the Pleistocene to create a coastal tundra environment. The Nunatsiavut coast is a lattice of ponds and lakes atop mostly bare rock veiled in dirt, kaleidoscopic lichens, and many different types of edible berries, each of which has an ideal month to be picked. This watery landscape gives way to a sea dotted with rocky islands of every size and shape, some erupting from the ocean toward the sky, and others appearing almost as accidents of the tide. It takes well over an hour by speedboat to get from Hopedale to the edges of the outer islands, where icebergs float past, sculpted by their ocean voyage from Greenland. This dangerous and complex Labradorean coast was deemed the "uttermost ends of the earth" by Moravian missionaries, colonizers who first arrived in the late eighteenth century to attempt to convert Inuit to Christianity. The colonizer and medical missionary Sir

Wilfred Grenfell begins his 1934 text *The Romance of Labrador* with a chapter entitled "The Pageant of the Rocks":

> The baseness of the rocks, their freedom from obscuring forest and turf, helps the long coast to tell its own geological story. Mother Nature has there taken off more than the usual amount of clothing which she is wont to bestow on the land elsewhere; and the autographed story of the ages is so imprinted on her naked bones that those who run may read its thrilling pages, and the wayfaring man can enjoy the conceit of being for a while a veritable Sherlock Holmes. To know Labrador is to know her geology. Seldom elsewhere is the explorer's mind so forced to think of the very beginning of things.[1]

The naked bones of bare rock, which impelled explorers to dwell on the prelapsarian nudity of the earth, reveal an awe and a fear of the way in which the place is rough, rugged, and unforgiving but also spatially and temporally expansive. Its geology inspires foreigners and residents to consider the deep past, its glaciation a shared object of contemplation.

Visiting Amos

The nudity of the earth, the near-complete treelessness of Hopedale, unlike other communities on the Labrador coast—all of which are more heavily forested—renders difficult the act of obtaining wood to burn. The geological as a contemporary condition of the landscape was something I was reminded of each time I walked into a woodstove-heated home from what can be a biting cold outside. In Hopedale, using a woodstove to heat your home is a significant task—it requires fuel, money, time, and knowledge. Warmth requires having something to burn. To tend the fire requires "going off" to the surrounding forested bays and islands south of the community. "Going off" in Nunatsiavut is a catchall phrase for leaving the community and traveling on the sea ice, ocean, or land to hunt, fish, collect water, visit cabins, or just spend time away from the main settlement.

Historically, Inuit were highly mobile, living either on the many bays that punctuate the coast or on the outer islands, depending on the season. Some scholars say that Nunatsiavummiut stopped living such mobile lives—that they settled—in the 1950s with the introduction of wage labor and the building of a Cold War–era American radar surveillance base in Hopedale. I argue, though, that critical mobility practices have never ended. Nunatsiavummiut still trace the outlines of the coast to the interior of the province and out to the edge, where outer islands get smaller and smaller and give way to the deep Labrador Sea. Many Nunatsiavummiut build cabins far away from Hopedale so that they can spend extended time on the land. The cabins are sometimes more than three hours away by snowmobile, each one kept warm by a woodstove, all sustenance cooked on the woodstove, life away from the glaciated community also propelled forward by the felling of spruce to burn.

The first fall that I was in town, I visited Amos's house for an interview. I had met him that summer at a lunch with his niece. We had talked over pea soup and locally picked redberry pie about his memories of the American radar base. He was in his late seventies and had moved to Hopedale when he was eight so that his father could find work on the base. During our interview he told me a story representative of "going off." He and his wife were at their cabin located far south of Hopedale, in a place that he describes as "right in the road" for snowmobiles on their way north for caribou. He and his wife had left Hopedale at dawn, arrived at their cabin in the late morning, and started up the woodstove. "I went for a load of wood, and she said she was going to clean up," he continued. "And when I got back—I was probably gone about an hour and a half—and I said, did anyone stop? And she said, yes, I gave out twenty-nine cups of tea while you were gone."

Amos's memory of his morning at the cabin is an account of the expected sociality of giving out tea to warm those who visit you, especially those headed north to find the caribou herds that feed on the lichens and mosses still present on rocky outcrops in the winter. The woodstove: stable and radiant, ready to turn bags of Tetley submerged in brook water into tea because of this social norm. Amos's first task upon arriving was to light the stove to ensure warmth; his second task was to fell trees to continue to ensure warmth. His cabin is a stop for those traveling to the glaciated, treeless north to find animals who make lives in subarctic ecologies, searching for complex lichens who also make lives in difficult subarctic ecologies. This memory reveals the primacy of keeping warm and the generosity at the heart of the social fabric of the circumpolar north.

Visiting someone to talk over tea in their home or cabin is the main form of socialization in Nunatsiavut. There are not many commercial enterprises—most of the communities don't have a restaurant and have at most one or two stores. This intensive visiting happens most nights a week—I have never felt less isolated or alone than when I am in Hopedale. Friends will stop by and drop off berries, duck eggs, arctic char, or partridges—gifts from the earth they collected while going off—or invite me over for supper. When heading to more unfamiliar houses, I knock on people's doors to announce my presence, but I am usually reminded that only police knock, and I should feel free to simply walk in. At each home I enter, my host will tend to the fire, doing the physical and mental act of determining when and how much wood to put in the stove to keep the home comfortable—an act of care. And the act of keeping the fire going is the only visible step for most visitors from outside of Nunatsiavut. But the walk from the mudroom to the woodstove with a spruce log in hand is the most minor in a chain of labor and knowledge.

As Amos hinted, keeping the fire going is much more complex in glaciated terrain, a bare-rocked earthen floor where there are no forests where people might find wood. The process is arduous, rooted in both place-based knowledge and physicality. Burning or building in Hopedale requires getting lumber from outside the community: from the forested bays, shipped up on the ferry, or from the back of someone else's snowmobile from a different community. Some Hopedalimiut travel three hours to a specific area where lightning struck one summer night in 1994, starting a forest fire that burned acres of tall spruce. The wood is already cured, so they don't need to wait for it to dry out. People still talk about going up the hill behind Hopedale and sitting on the remains of the radar base to watch the flames lick the forest in the far-off bay. Obtaining wood in a community with little soil requires an intimate knowledge of place, climate (including weather), and temporal commitments.

Nathan and Practices of Care

Treelessness due to glaciation as an environmental particularity was voiced in many interviews with Hopedalimiut when asked about what sets their community apart, as Hopedale is now the only remaining Nunatsiavut community with this type of bare-earthed condition. Other northern Nunatsiavut communities with glaciated landscapes were forcibly relocated as recently as the late 1950s—a reminder that settler colonialism is not just an artifact of the past. In Hopedale, those who are unable to cut down their own wood must buy or get wood from neighbors. Oftentimes families look out for one another, or elders are given wood by younger men they are close with.

Last spring, I accompanied a young man who went wooding on behalf of a woman whose husband had just passed. He had already chopped down several trees by his cabin, a cozy two-room home painted bright red on a forested island twenty minutes from town. We threw the splits he had made into his *kamutik* and drove back home on his snowmobile as the sun was setting, both of us sweaty under our parkas from the work. On a separate occasion, I helped a middle-aged woman named Maggie bring wood into a different widow's basement piece by piece, stacking it up over the course of a long afternoon. The sociality of warmth—the labor of keeping the space inviting, the dynamism of controlling your own heat, the intimacy of cutting down the wood for someone else—is embodied and intensive.

One November evening in Hopedale, I was interviewing an elder named Nathan whose wife had urged me to call on him to discuss his memories of his childhood in the community. I set up the woodstove on my own, fiddling with the knobs that control airflow and throwing in one too many birch logs. It burned too hot as we talked, and he took notice, using that moment to share with me that he remembered walking up the road to the radar base as a child after it had been abandoned by the US Air Force and left to decay. His family and others who could not afford gas or to keep a dog team to obtain wood from outside Hopedale would use axes and hand saws to chop down the huge telephone poles that flanked a winding, American-made road that snakes up to the cement foundations atop the gneiss hills. The telephone poles were burned for warmth. The remains of them are still there, creosote-soaked trunks too large to be local, stuck into piles of man-made blasted rock, ax marks visible. These huge pieces of chemically treated lumber were brought north during the 1950s during the base's initial construction, transported by sea on military supply ships.

The hidden work and histories of collecting wood are hidden only to those who are not from Hopedale. This narrative excavation of a woodstove illuminates the way that earthen specificity is constitutive of people and place. Cold places, places with rocky shores and ancient geologies, are also places of immense care. Keeping warm, providing warmth for others, is a form of care with geological grounding. Glaciation makes this place, makes it cold, and makes it difficult to keep warm. The labor of Nunatsiavummiut circumvents this in both spectacular and intimate ways. Social warmth—as labor, as joy, as a complex relation—is a way to consider our own attunement to Earth as place, as planet, as material. Each conversation with a Hopedalimiut knowledge keeper is a glimpse into how warmth is cultivated, how it is not taken for granted.

Warmth requires something to be burned. I think about this when I am not in Hopedale too, when I am in my urban home, where networks of labor can be difficult to trace, where the earthen elements of warmth can seem immaterial or impersonal, where they are often concealed. Grabbing a splintery piece of wood from the mudroom to toss in the woodstove—from a tree that grew in faraway forests, that was brought to the community by snowmobile over the frozen sea, and that dried out on the side of the dusty road for the better part of the year—is a material reminder of what the earth can provide, if one does the work to attune to it.

note

1. Wilfred Thomason Grenfell, *The Romance of Labrador* (New York: Macmillan, 1934), 1.

Border Crossing, Kentucky

Melissa Tuckey

The light on your hills is sadness
that lowers into dusk

I do not speak your language
nor have I trained
my ear not to miss you

I've known your ghosts
lived next to strangers
watched the old man and his mules tend
to slowness in the landscape

Heard tell Made do
Broke something
trying to fix it

I've seen the mountain
blown clear off to make
some kind of progress
out of nothing

Been as flat inside as a landscape
after the rush for coal

Watched the water turn brown
and the snakes come up

Lurked awhile in the whippoorwill dark
clinging to every sound

And still, the brightness shines through the holes in the earth

Imani Jacqueline Brown

*T*he air is different here, my partner, Matthieu, remarks upon exiting the airport. Matthieu's first visit to New Orleans is like a first step onto the moon.

Yes, there's a heaviness to the air; it's a result of the humidity, I answer easily.

Yes, but also something else...

Oh, it's the emissions from the petrochemical plants that surround the city, I offer readily. Their noxious odor is more pervasive than ever. It's even hanging over the wealthy Garden District neighborhood, and the city's most privileged residents are outraged. My hypersensitive eyes and skin start burning not long after we leave the airport.

Yes, there's that, but there's something else still, something otherwise. Something like... dancing.

Oh, you mean the brightness!

I was raised in New Orleans, Louisiana—spent my childhood there until 2005, when Hurricane Katrina left 80 percent of the city under water, displacing my family along with one hundred thousand other Black residents. The city's "rebirth" has taken the form of gentrified replacement. These days, I oscillate between states of wanderlust and nostalgia, unsure of the prospect of "homecoming" to a place promised a future of climate displacement.

Desiring security, I attempt to settle in London; unable to stay away, I settle my gaze upon Louisiana's image like a distant lover.

I dedicate my life to mapping the forces that displaced her people into a diaspora and to sharing the story of her destruction with the world. Through the newly remote perspective of satellite imagery on my laptop, I learn to sense Louisiana's topography as an archipelagic extension of my own body, to read the oil wells and canals in our wetlands as open wounds in our ecological skin. But how can I convey a sense of the spirit that continues to draw me to her body?

I was surprised while reading Jeff VanderMeer's *Southern Reach Trilogy* to stumble upon the perfect language to describe that ineffable, centripetal force. VanderMeer's books, situated in an indeterminate wetland geography called the Southern Reach, begin untold years after a succession of military and scientific expeditions have tested the impotence of guns and theories against an unknowable, more-than-human force known only as *the brightness*. Touched by the brightness, species mutate: humans become animals becoming plants becoming frightening specimens of an uncanny in-between. One by one, missions fail and succumb; the brightness spreads ferality through the military's containment zone, from coastal salt marsh through brackish swamp and right up to the bottomland hardwood border beyond which the military's base squats. But if the borders erected between species couldn't hold up, then certainly the military's self-sequestration wouldn't stand a chance. Acceptance (aptly the title of the third book in the trilogy) of Earth's ability to wrest control of the levers of change was the inevitable conclusion from the first.

And so, I began to speak of that *something else* between Louisiana's heat and humidity, poltergeists and pollution, as the brightness that pervades the fictional swamplands of the Southern Reach.

1. The brightness is the power to become otherwise

Swamps everywhere are enmeshed in cycles of becoming otherwise: life becomes death and death begets life, soil and water ebb and flow, and everything mixes in the murk. We humans glimpse the complexity of deltaic systems only through our personal points of intersection—the moments where our bodies meet the earth. But if we stretch out our consciousness, we can sense much farther than the reach of our fingers.

There's a reason marshlands born of riverine deltas from Egypt to Iraq to Nigeria to Louisiana are cradles of biodiversity and human civilizations. Deltaic ecosystems are *great attractors*, drawing human and nonhuman beings into their gravitational fields.[1]

There's magic in those swamplands. Magic and horror.

As wetlands attracted Indigenous civilizations, so did they attract those civilizations' destroyers: colonizers, slave traders, sport hunters, oil corporations.

Churning vortexes of life and death settle, layer, and accumulate as vast subterranean fields of oil and gas, which we might think of as the final resting places for millennia of life.

I recognize Louisiana's reflection in the shifting landscape of the Southern Reach because in Louisiana, an external power has disrupted her ancient systems and imposed its own, tipping the scale toward her (and our) becoming otherwise. That power is colonialism. Its system is extractivism.

Extractivism deploys the force of segregation to divide existence into alienable, fungible, divisible, accumulable fragments. Extractivism declares Indigenous lands empty, prefiguring their genocidal dispossession. It alienates human beings from our ecological bodies and racializes Black bodies beyond the borders of the body of humanity.[2] It corsets the Mississippi River with levees, segregating river from land and impeding land-generating cycles of floods and alluvial deposition. It exiles Black, Brown, and Indigenous communities to the most precarious geographies and disperses us to the eroding edges of the Earth.

Untold millions of holes are drilled ten thousand feet deep into the body of Earth. Millions of barrels of oil gush forth into our waters, for five months here and nineteen years there.[3] Corporations deploy chemical dispersants to sink slicks out of sight and mind; their emissions lurk beyond the threshold of visibility, lingering among clouds, settling in cells. Air and water are toxified, mutating the building blocks of life.[4] Two thousand square miles of coastal wetlands erode; ecologies unravel. The air shifts, the currents fail, species and places once familiar become strange.

I recognize Louisiana's reflection because somehow, in spite of it all (or perhaps because of it?), a *brightness* shines through holes drilled through the fabric of existence.

Our ancient, nonhuman ancestors, says Malidoma Patrice Somé, a shaman from the Dagara community of Burkina Faso, *dwell beneath the surface of the earth, forming a vast pool of energy.*[5]

Malidoma was a dear friend of my mother's. I had hoped to study under him, to learn his perspectives on ecological philosophy, but he passed unexpectedly in 2021. These days, I imagine us in dialogue across the Bakongo Kalûnga line, across the ocean of death and rebirth, through lines of text in his books.[6]

Oh! Is theirs the energy that we pump and burn? I ask Malidoma. *Is climate change the exhalation of the immeasurable sighs of our primordial kin at the intransigence of racial capitalism in its destruction of the planet?*

Is theirs an exhaust(ion) too grave to be tabulated in parts per million?

As Louisiana disintegrates, my body swells with grief. My face swells with unshed tears. My neck, my shoulders, my back, my arms, my chest, my belly twist and warp, contorted and concave, to shield my heart from the retreating embrace of the world. My muscles grow around my grief, connective tissue calcifying to bone; my skin rashes, revealing my inner turmoil. I am a system out of balance.

I notice how grief territorializes and mutates the geography of my body and wonder how nonhuman bodies spatialize their grief.

Nature is the only place that can hold the spirits that leave the cities and towns, remarks Malidoma. *It is beyond thought what will happen if nature, already so endangered, is destroyed.*[7]

The shipwreck of our souls under racial capitalism is externalized as landscape. As Louisiana unravels, I unravel. It is undeniable that my move to London was a futile attempt to flee this unraveling.

I would not use the word exile *for myself*, I say to the Lebanese filmmaker at the winter barbecue in East London, *but I feel as though I can never live there again, and I feel that grief in my bones.*

That feeling in your bones, replies Malidoma, *is knowledge in the form of memory.*

My body remembers every moment, feels every change, holds every loss from a disaster three hundred years old and still unfolding.

It drives my partner mad that I can't let go of the past.

2. The loss of my motherland is to the earth the loss of a child

Before colonists enslaved her and erased her given name, Louisiana was known as Bulbancha, meaning "land of many languages" in Mobilian, a unifying trade dialect of Choctaw. The vibrations of shared language wove together over two dozen nations:

Atakapa, Avoyel, Bayogoula, Biloxi,
Chatot, Chawasha, Chitimacha, Choctaw, Houma, Koasati,
Koroa, Mugulasha, Muskogee, Natchez, Natchitoches,
Olelousa, Opelousa, Pascagoula, Quapaw, Quinipissa, Souchitioni,
Taensa, Tangipahoa, Tawasa, Washa

The Mobilian language speaks to the heart of Louisiana. It reminds us that she is a place of convergence, integration, and flow.

At a mere seven thousand years old, Louisiana's land is young. The loss of my motherland is to the earth the loss of a child. As alluvial soil, Louisiana's land has youthful energy. She wants to be revitalized and refreshed, to mingle with new sediment traveling down the Mississippi River from faraway places. Our land is as loose and free as our hips and our spirits. When we feel ourselves, we flow with our wider ecologies.

Yet for three hundred years, colonialism has forced us to settle, to be stiff and still, to resist the Delta's desire to flow with change. Is this why we humans feel so stuck? Stuck in time, stuck in our ways, stuck in the abyss of racism and corruption and evasion and denial...

And yet as we witness our land and our people slipping away, our tongues are let loose.

Across Louisiana, in Indigenous communities from Isle de Jean Charles to Houma, and in Black freetown communities from Ironton to Welcome, resistance erupts like levee breaks, flooding the earth with new energy and fresh ideas so that life can branch rhizomatically.

A couple of years ago in London, I initiated and led an investigation with Forensic Architecture, a human rights research agency, into environmental racism in a Louisiana region once called "Plantation Country," and now known as the "Petrochemical Corridor," nicknamed "Cancer Alley."[8] Rise St. James, a grassroots activist group from the community of Welcome, asked us to develop a methodology for locating the burial grounds of historically enslaved people before new industrial complexes broke ground. To determine the most likely locations of those cemeteries, we needed to understand the spatial logics of antebellum sugarcane plantations.

As part of our research, I guided the creation of an immersive computer model of the "back of the plantation"—the underresearched realm where enslaved Black people lived, labored, and were buried. Our goal was to bring atmosphere and depth to historical maps and photographs, not to produce a one-to-one replication. Still, I was dismayed by the impossibility of an accurate rendering of the plantation's oak trees.

Louisiana's oak trees are singular. *They should be heavier, thicker, more expressive,* I told our computer artist. Their energy is languorous and steady, slow yet responsive to change. Make them thick like the swamp air, thick like the swamp mud, thick like roux, thick like thighs, thick like time. Their branches make a start at growing upward but seem to abandon the effort to conform to an idealized shape. Why reach stoically to the sky when one can stretch out long long, reaching for other oaks across the way, resting one's heavy limbs against the ground?

Our trees know the length of the struggle and are in it for the long arc. Our trees look like they have lived. Lived *through* things. Seen things. Witnessed things. Survived things. They should look like they have borne the weight of history, borne the weight of the world, borne the weight of Black bodies swinging from their thick, twisted branches for hundreds of years.

I would never have dreamed of 3-D modeling human beings, a self-evident hubris. But the rest of the lesson I had to learn in the act: Whether human or nonhuman, a body is an expression of a life lived. No reproduction can capture the spirit of those trees.

The trees here are unbelievable! Visiting New Orleans in January, Matthieu stops before each tree we pass, caressing its textured bark, imagining the flowers it will bear come springtime, wondering at its name: Southern magnolia, bald cypress, crepe myrtle, weeping willow, live oak.

He gathers a handful of acorns with all the earnestness of a wintering squirrel. *We should take some superresilient Louisiana acorns and plant them in London.*

Yet my friends in Louisiana remark that our oak trees seem sick. Their leaves are thinning out; their crowns can't reach their kin across the way. Is it all too much for them?

3. *Mourning what we have been frees us to rejoice at what we might become*

Matthieu and I separate. I spend February by the English seaside, alone. I fall apart and come together again.

Throughout it all, the earth is here for me. Yemaya, spirit of the Atlantic Ocean, reminds me that England's waters are the same as the waters of the Gulf of Mexico. With her, I am always home. How do I show up for her? What gift could I possibly offer in thanks? In what language would she comprehend my gratitude?

Malidoma tells me that, for the Dagara, *to utter* signifies nostalgia for humanity's true home, which is in nature. Expression through poetry points to our exile from nature, expresses our innocent and imperfect yearning to embrace the ineffable wonder of Earth.[9]

I shed words like dead leaves, releasing them to become humus, the matter of infinite return. My grief releases in wails, cries, songs, plumes of carbon dioxide. Trees capture my emotional exhaust and recycle it, photosynthesize it into glucose for their own sustenance and oxygen for my own. I inhale great, sobbing gulps of it thirstily, gratefully, joyfully.

Across Louisiana, chants and speeches and legal arguments rise to the trees. They demand a moratorium on petrochemical development, demand accountability for centuries of racial violence, demand the recovery of our erased ancestral heritage.

In 2022, a Louisiana district judge issued a legal decision that lifted up the words of Sharon Lavigne, founding organizer of Rise St. James, reaffirming her appeal to the sacred quality of the land holding burial grounds of historically enslaved people on Louisiana's petrochemical plantations.

But where does sacred land begin and end? Sacred land is seeded by human hands that plant the trees that hold the land together. It spreads through roots and branches, entwining and migrating with birds, insects, and fungi. It is exhaled by leaves into atmosphere.

Sacred land does not end at the plantation property line, where earth is dug out into industrial pits and filled with petrochemical waste. It does not end at the expanding horizon of open water as fragments of earth collapse into the sea. It does not end at the human bodies who defend their right to live on their land until the ends of the earth.

We are all the ground whose vegetables we eat. We are all the trees whose air we breathe. We are all the oil we spill and the chemical dispersants we spray. We absorb and become the places we are in. We are the earth we inhabit.

By the end of the *Southern Reach Trilogy*, one is left wondering whether the brightness emanates from the uncontrollable impact of an external force or whether it represents the adaptive response of Earth to such a power. Wherever it comes from, humans absorb it, become it, become changed by it. And life goes on.

Saidiya Hartman asks whether five long centuries of systematized racial brutality were prefigured by the first arrivals of solitary Portuguese traders to the shores of West Africa. Did the future begin to unravel from the point at which the first Portuguese flag staked the earth?

The holes staked by colonial flags have spread and mutated into vast expanses of pockmarked oil fields and eroding wetlands. Wherever we thought we were going, we're here now. We know now.

We have been changed by the forces of extractivism unleashed upon the earth, and now we must guide that change: we must transition. The brightness shines through the holes in our souls, as through the holes in the earth. As we mourn and release what we have been, we free ourselves to rejoice at what else in Earth's image we might become.

notes

1. A gravitational attractor is a region of space with a massive concentration of galaxies. This high concentration of mass exerts a tremendous gravitational force that pulls astral bodies together, counteracting the force of expansion that seems to pull bodies in the universe apart. The Great Attractor is one such region containing the Milky Way along with one hundred thousand other galaxies.
2. See, e.g., Imani Jacqueline Brown, "Ecological Witnessing," in *Fieldwork for Future Ecologies*, ed. Bridget Crone, Sam Nightingale, and Polly Stanton (Eindhoven, The Netherlands: Onomatopee, 2023); Naomi Klein, *This Changes Everything: Capitalism vs. the Climate* (New York: Simon and Schuster, 2015).
3. The BP Deepwater Horizon oil spill began with an explosion that killed eleven workers on April 20, 2010; the rig gushed oil until September 19, 2010. A little-known spill from one of Taylor Oil Company's offshore wells has leaked unabated, spilling over three million barrels of oil since 2004, when Hurricane Ivan destroyed an oil platform in the Gulf of Mexico.
4. Oliver Milman, "Deepwater Horizon Disaster Altered Building Blocks of Ocean Life," *The Guardian*, June 28, 2018, https://www.theguardian.com/environment/2018/jun/28/bp-deepwater-horizon-oil-spill-report/.
5. Malidoma Patrice Somé, *The Healing Wisdom of Africa: Finding Life Purpose through Nature, Ritual, and Community* (New York: Tarcher, 1999), 190.
6. Zion Murphy, *Past, Present, and Future Cosmically Intertwine. | The Kalunga Line*, YouTube video, 4:42, posted by Zion Murphy, January 31, 2023, https://www.youtube.com/watch?v=6UY-7DODtx4.
7. Somé, *Healing Wisdom of Africa*, 54.
8. Forensic Architecture, *Environmental Racism in Death Alley, Louisiana*, 2021, https://forensic-architecture.org/investigation/environmental-racism-in-death-alley-louisiana/.
9. Somé, *Healing Wisdom of Africa*, 49–50.

With Mouths and Mushrooms, the Earth Will Accept Our Apology

Franny Choi

*When Hiroshima was destroyed by an atomic bomb in 1945, it is
said, the first living thing to emerge from the blasted landscape
was a matsutake mushroom.*

> —Anna Lowenhaupt Tsing, *The Mushroom at the End
> of the World*

made of shock doctrine, made of root-chatter,
made of playgrounds blackened with corpses,
made of the anti-modern, of hell-wreck, made of symbiotic
destruction, of parasite and pericapitalism, made of slave trade,
of wretched, made of reek and reason, made of ex-flesh, of cells
stampeding through the lungs of miners, made of morals,
        of mortals,
of *everything went with perfection*, made of infants in horse stalls
or strapped in reddened rags, made of sour rice, of phlegm,

of warp and spore, made of forest floor dis-
enfranchised to dust, of sun-laden and empty space, made
of empty space, made of timber price, of sugarcane,
of nematode and elite sleaze, of seamstresses diving to concrete,
made of trash-bellied caribou, of suicide pact, of hands deep
in the throat of a comfort girl, made of entrails, of gromet,
of every chipper *we can do it,* made of sacred land,
    of funeral scream,
of holy water and flayed graves, made of stolen names,
    of commerce,
of conquistador, of near-death visions on the factory floor,
    made of
the unforgivable future, and the unforgivable past, i bloom,
bloodless, and ready to feed.

The Medieval, the Modern, and the Mundane

Danielle B. Joyner

My parents met and married in Salt Lake City before moving away to start their lives together. In subsequent years, when our summer vacations were spent in this valley ringed by mountains and desert, I believed that nothing was more beautiful than the view of Mount Olympus from my grandparents' backyard. We returned to Salt Lake City when I was ten years old and moved into the East Millcreek neighborhood, along the foothills of the Wasatch Front, where, to my delight, I had an even closer view of my favorite rock-topped peak.

The dramatic landscape surrounding this western city shaped my coming of age, from hiking up canyons and skiing down the slopes at Alta Ski Area with my family to my first summer away from home, when I worked at the Girl Scouts' Camp Cloud Rim, near Guardsman's Pass above Park City. In high school and then at the University of Utah, my adventures spread farther afield to include the western desert and the magical lands of central and southern Utah. After studying medieval art history at the University of Utah, graduate school and subsequent employment led me eastward onto flatter, more subdued terrains. Professionally, I am now an art historian and a medievalist; personally, my character was indelibly shaped by the varied terrains of the mountain west. It is a special treat, then, to combine both elements of my professional and personal self in an essay that considers convergences of the medieval, the modern, and the mundane.

Medieval refers to the European Middle Ages, that poorly named millennium between the fall of the Roman Empire (c. 450) and the gradual emergence of what we label as early modernity— that is, the Renaissance (c. 1450). This era might not immediately suggest itself as relevant for twenty-first-century concerns about the environment. I contend, however, that these long-ago centuries offer fascinating comparisons that enrich how we think about our own actions and contextualize some of the pressing issues of our times. Although the term *modern* generally refers to the historical era from the seventeenth century to contemporary times, I use it more loosely to refer to activities that I have enjoyed during my own lifetime, in this case, hiking up mountains and collecting stones and earthen bits from different places I have visited. By *mundane*, I do not mean dull or ordinary. Rather, I use the term in light of its Latin root, *mundus*, to mean "the world," or more specifically, "the earth beneath our feet."

Climbing

Mountains beckoned to me throughout my teenage and early adult years. Among many enjoyable excursions up canyons and across ranges, the moments most deeply etched in my being are those ascents that I undertook on my own. Whether to wrangle with pubescent turmoil or simply to take advantage of a midwinter thaw, scaling the hillsides above the city offered soothing respite. The views afforded by a high perch, both across the flat valley and along serrated peaks, instilled in me a sense of wonder at the Earth's magnitude, humility for my own transitory nature, and hope for our combined future. It is not unreasonable to suggest that a similar trio of wonder, humility, and hope also inflected chapels built during the Middle Ages in honor of St. Michael.

As one of the four archangels, Michael was characterized by medieval Christians as the powerful warrior who cast Lucifer out of heaven at the beginning of time and who, according to the New

Testament's book of Revelation, will defeat the evil dragon at the end of time.[1] His angelic status, however, did not prevent believers from creating for him a cult emulating those of canonized saints, with shrines and descriptive vitae (biographical lives of the saints) recounting his miraculous deeds. An eighth- or ninth-century vita titled *De apparitione Sancti Michaelis* (On the Apparition of St. Michael), recounts how the archangel made his first appearance in Western Europe in 492 CE on Mount Gargano, where he miraculously protected a bull from a disgruntled landowner.[2] In subsequent visits to this site, which is in modern-day Apulia, Italy, Michael also assisted local Christians in their battles against pockets of die-hard pagans, and then he marked a spot inside a mountain cave with "the footprints of a man, as it seemed, firmly impressed in the marble," to designate where he wanted a shrine in his honor to be placed.[3] The local bishop, who had been unsure of how to proceed, received a helpful vision in which Michael appeared and said: "It is not your work to dedicate the church which I built. For I, who built it, also dedicated it myself."[4] This cave in the mountains became a consecrated chapel, and as the cult of St. Michael gained renown, numerous Christian pilgrims visited the blessed cave to pray for the archangel's intercession and support. To this day, the Sanctuary of Monte Sant'Angelo on Mount Gargano remains a popular destination among pilgrims and tourists alike.

A pilgrimage to Apulia was one means of honoring St. Michael, and another was to build shrines, churches, and even monastic foundations dedicated to him. As adoration of the winged saint spread northward, the devout kept in mind Michael's pronounced preference for lofty, dramatic, and isolated locales. By building in terrains that resembled his chosen mountain cave, medieval Christians hoped to curry Michael's favor and bestow on their locale the sanctity associated with his original fifth-century shrine.[5] Two well-known examples of this are Skellig Michael, a steeply rugged island off the western coast of Ireland where a monastic foundation was established in the sixth century, and Mont Saint-Michel, a tidal island on the coast of Normandy where another monastery dedicated to the saint was founded in the eighth century. Among other examples of buildings and foundations dedicated to this well-loved saint, it is the small tenth-century Chapel of St. Michel d'Aiguilhe in the Auvergne region of southern France that, much like the rocky peaks of Mount Olympus in Salt Lake City during my childhood, captured my attention (fig. 1).

Figure 1. Chapel of St. Michel d'Aiguilhe, Le-Puy-en-Velay, France. Photo by Danielle B. Joyner.

In the town of Le-Puy-en-Velay, a narrow basalt volcanic neck rises 272 feet from the valley floor.[6] On July 18, 961, Deacon Truannus received permission from Bishop Godescalc of Le Puy to build a chapel dedicated to the archangel Michael on the peak of this steep, needlelike neck.[7] Given the obvious difficulties in scaling the nearly vertical basalt while hauling up the tools and stone necessary to build the chapel, the perilous location might seem an impractical choice to a twenty-first-century mind. To the mind of a tenth-century Christian, though, especially one who had endured several decades of increasing violence in the region, it was the perfect spot for a chapel dedicated to an angel-saint renowned for their protection, for their military prowess, and especially for their avowed appreciation of high, isolated perches.

Medieval Christians believed in a world where angels waged constant battle against evil and where chapels dedicated to saints, or angels, endowed their physical settings with palpable sanctity and power. High above the town, the chapel dedicated to the archangel Michael would have been a source of wondrous protection as well as a constant admonition for humility—after all, it was Lucifer's sin of pride that caused him to be cast out of heaven by Michael. The shrine-topped peak also would have been a source of hope, not only for the townspeople but also for the many travelers who made their way through Le Puy. Bishop Godescalc, mentioned earlier, was one of the first recorded French pilgrims to visit the shrine of the apostle St. James at Santiago de Compostela in Galicia, now modern-day Spain.[8] His trip occurred in 950–951, and it was surely not a coincidence that the seat of his bishopric became the starting point for one of the main pilgrimage routes across the Pyrenees to the apostolic shrine at Compostela. Indeed, so many visitors were climbing the more than 230 steps carved into the steep volcanic neck that the chapel was enlarged in the twelfth century to accommodate their numbers. Imagine how the striking view of the archangel's lofty shrine would have encouraged medieval pilgrims to walk a bit farther that day while inspiring hope in them for the rewards at the end of the long path ahead.[9]

Collecting

As an art historian, I am drawn to beauty, to fabrications that reveal the skill and thoughtfulness of their creators, and to objects that are—I am not ashamed to say—pretty and sparkly. This fascination with and love of visual splendor had a different focus in my early years. As a child, especially colorful pebbles glistening beneath mountain streams called out to me, and after being plucked from cold waters, they often ended up in jars and egg cartons on my bookshelves. As a young adult, oddly shaped stones rising from sandy trails in the western desert similarly demanded my attention. To this day, a heavy rock formed of three spheres adhered together in a lopsided tower still stands on my mantle as a lithic reminder of life before graduate school. A variation of this theme also resides on my refrigerator door in the form of a magnet filled with sand from an island that I had the good fortune to visit. Even in the coldest winters, this inexpensive token evokes sensations of sun-warmed sand between my toes. These acquisitive actions, which a cynic might characterize as childish habit, sentimental whimsy, and blatant consumerism, acquire new valences when compared with choices of devout Christians living in the early Middle Ages.

In the late sixth century or early seventh century, a Christian pilgrim visiting Jerusalem and the Holy Land picked up stones, sticks, and other earthen bits from sites associated with significant events in Christ's life.[10] Inscriptions written on some of the small rocks identify their origins, such as "from the life-giving [site of the] Resurrection."[11] Set into a bedding of hardened sand, this collection of stones survives in the Lateran Palace in Rome in a painted wooden box that was likely made in Palestine around the year 600 CE (fig. 2). The locations represented by the earthen bits reappear on the inside of the lid as five colorful paintings portraying seminal moments from Christ's life: the Nativity, Christ's baptism in the river Jordon, the Crucifixion, the empty tomb, and the gift of the Holy Spirit to the Virgin Mary and the apostles at Pentecost.

Contrary to their nondescript appearance, these earthen bits were believed to be impressed with the divine. The modern label "contact relics" refers to objects that have been in close, physical proximity to a saint and therefore have absorbed the saint's divinity. The earthen bits in this box were collected and treasured not only as mementos of a journey to Jerusalem but also as actual pieces of the Holy Land. They bore witness to the most important moments

Figure 2. Wooden Reliquary, Syria or Palestine, 6th century, Lateran Palace, Rome. © Photo copyright Governate of the Vatican City State-Directorate of the Vatican Museums. Used with permission.

in Christ's life and perhaps were even trod upon by Christ, his mother the Virgin Mary, or one of the twelve Apostles. Simply put, Christ's divinity was instilled within them. Thus, when the pilgrim returned home from their travels, they could open their box to contemplate souvenirs of the Holy Land in concert with imagery depicting the events that impressed sacrality upon the very stones. Sharing in the beauty of the painted imagery, these earthen bits would have inspired memory to intertwine with prayer.

A variation of these stone relics are the earthen tokens obtained by pilgrims to the shrine of St. Simeon the Stylite, near modern-day Aleppo, Syria (fig. 3). When Theodoret, bishop of Cyrrhus, wrote the first life of St. Simeon in the mid-fifth century, the stylite saint was still living.[12] Theodoret recounts that Simeon was born in the late fourth century in a northern Syrian village and as a child worked as a shepherd. One day he went to the temple, and after hearing the Gospel message, he joined a group of local ascetics and began a life of physical devotion so extreme that he was asked to leave the community. Not only did Simeon fast constantly, but he wound a rope so tightly around his body that his flesh rotted and attracted maggots. Then, for extended periods of time, he occupied a dirt hole dug in the garden. Considering these intense bodily afflictions, festering sores, and what must have been eye-burning stenches, the community's aversion to Simeon is understandable.

Figure 3. Pilgrim token from the Shrine of St. Simeon, sixth century. Baltimore, Walters Art Museum. Creative Commons License.

Nevertheless, word of his ascetic devotion spread, and people flocked to see a person they believed was already a saint. Theodoret notes that it was the constant presence of pilgrims reaching out to touch his seeping flesh and crusty robes that drove Simeon to ascend and live atop the first of three successively taller columns. It was this practice of residing atop columns that lent him the sobriquet of Stylites, from the Greek *stylos*, for "column." Simeon's final lofty abode rose approximately sixty feet into the air and supported a six-foot "basket" to accommodate his multiyear occupation. Reaching an impressive seventy years of age, Simeon's death and burial further sanctified the site, upon which a monastery was built to accommodate pilgrims and additional stylites adherents.

In the following centuries, monks overseeing the holy site would collect dirt from it, mix water or oil into the dirt, and then create round tokens impressed with images of Simeon or Christ.[13] Fired into a durable, terra-cotta-like consistency, pilgrims purchased the tokens as sacred relics, souvenirs that carried Simeon's hard-earned sanctity back to their homes. The miraculous efficacy of the tokens was asserted in a second account of the life of St. Simeon, which likely was written by his disciples shortly after his death. Not only was it believed that the sacred dirt of the tokens could heal chronic illnesses in people, but various testimonies claimed that pieces of the tokens buried in the earth revived long-dead gardens or kept dangerous wild animals at bay, and one token even exorcised a demon from a storm-besieged ship of pilgrims. The pilgrim tokens, just like the stones, sticks, and sands collected in the Holy Land, speak to the beliefs of early Christians as well as to the travails of their world.

Conclusion

What can be learned from these medieval and modern reactions to the mundane? Perhaps most important, the similarities between actions separated by centuries and continents are striking. There is a shared compulsion to climb mountains, hills, and peaks, yet the motivations for climbing or the emotions elicited by their heights are expressed in an historically dependent manner. In the late twentieth century, a mountainside perch with a view elicited in me a powerful combination of wonder, humility, and hope. In the tenth century, a lofty peak was considered appropriate for a shrine dedicated to an archangel who inspired wonder, had no patience with pride, and who could be called upon for protection. In a similar fashion, despite the intervening centuries, there exists a desire to pick up and carry home bits and pieces of the earth.

I characterize my compulsion as combining a love of the beautiful with my need for physical reminders of significant times and places in my life. Today, I look at the oddly shaped stone on my mantel and remember how heavy my backpack became as I carried it back to the truck. This trilobed rock, a talisman of sorts, sat on two different desks as I wrote first my dissertation and then a book. It graced my window ledge during an unexpected turn in Dallas, and now it resides on my fireplace, stalwart and constant. How different, really, is my commitment to this rock from what a pilgrim might have felt, gathering stones from the ground on Mount Calvary, where Christ was said to be crucified, and carrying them home to be treasured in a beautifully painted box?

Maybe the pilgrim who purchased the token of St. Simeon Stylites ran their fingers over its hard, decorated surface and smiled at the memory of the journey to the site of this holy man just as I look at the sands in the refrigerator magnet and smile at my memories of an island idyll. Genesis 2:7 describes how God created the first man from exceptionally humble, mundane origins, "And the Lord God formed man of the slime of the earth," and Genesis 3:19 reasserts this elemental connection between humanity and the earth when God reprimands Adam and Eve for their disobedience in eating the forbidden fruit: "In the sweat of thy face shalt thou eat bread till thou return to the earth, out of which thou wast taken: for dust thou art, and into dust thou shalt return."

I do not need to believe the literal sense of these Judeo-Christian verses to acknowledge their deeper truth. Whether we consider ourselves medieval or modern, we are bound to the earth. Perhaps we modern folk would do well to remind ourselves of medieval responses to the earth and emulate their belief in the power and sanctity of the mundane—the *mundus*—that is, the earth beneath our feet.

notes

1. Apocalypse of St. John (Revelation) 12:7–9: "And there was a great battle in heaven, Michael and his angels fought with the dragon, and the dragon fought and his

angels: And they prevailed not, neither was their place found any more in heaven. And that great dragon was cast out, that old serpent, who is called the devil and Satan, who seduceth the whole world; and he was cast unto the earth, and his angels were thrown down with him." All biblical quotations are taken from the Douay-Rheims Bible Online (https://drbo.org).

2. For an English translation of this Latin text, see Richard F. Johnson, *Saint Michael the Archangel in Medieval English Legend* (Woodbridge, UK: Boydell Press, 2005), 111–15.
3. Johnson, 113.
4. Johnson, 113 and 115.
5. See also John Charles Arnold, *Footprints of Michael the Archangel: The Formation and Diffusion of a Saintly Cult, c. 300–800* (New York: Palgrave Macmillan, 2013).
6. I would like to thank Marcia Bjornerud, Walter Schober Professor of Environmental Studies and Professor of Geosciences at Lawrence University, for her explanations of this type of formation.
7. A charter of the foundation is preserved in the *Gallia Christiana* and reprinted in Xavier Barral i Altet, "La Chapelle Saint-Michel d'Aiguilhe au Puy," *Congrès archéologique de France* 133 (1975): 230–313.
8. Roger E. Reynolds, "A Precious Ancient Souvenir Given to the First Pilgrim to Santiago de Compostela," *Peregrinations: Journal of Medieval Art and Architecture* 4 (2014): 1–30.
9. See also Jessica Streit, "Pilgrimage and Liminality at the Cathedral of Notre-Dame, Le Puy-en-Velay, and the Saint Michael Chapel, Aiguilhe," *Notes in the History of Art* 31 (2019): 6–16.
10. "Wooden Reliquary Painted with Scenes from the Life of Christ," Musei Vaticani, https://www.museivaticani.va/content/museivaticani/en/collezioni/musei/cappella-di-san-pietro-martire/reliquiario-in-legno-dipinto-con-scene-della-vita-di-cristo.html.
11. Elina Gertsman and Asa Simon Mittman, "Rocks of Jerusalem: Bringing the Holy Land Home," in *Natural Materials of the Holy Land and the Visual Translation of Place*, ed. Renana Bartal, Neta Bodner, and Bianca Kühnel (Abingdon, Oxon: Routledge, 2017), 160.
12. *The Lives of Simeon Stylites*, trans. Robert Doran (Kalamazoo, MI: Cistercian Publications, 1992).
13. Gary Vikan, *Early Byzantine Pilgrimage Art* (Washington, DC: Dumbarton Oaks Research Library and Collections, 2011).

Is Paddy Heneghan Dead?

Liam Heneghan

For all things are from the earth and
all return to the earth in the end.
—Theodoretus, from *Treatment of the Greek Conditions*

Earth to Earth

1. Before my father's final illness, I had not understood that
    when the call goes out to assemble family members dolefully
    about a deathbed, it's because a decision has been made to let
    the loved one pass. The timer is set; the sand is trickling down.
    Death—that omnipresent possibility at the best of times—
    becomes calculable when sustenance is withheld. Paddy
    Heneghan, born in Tralee, Ireland, on 29 March 1927, lived
    ninety-five years and, though defying predictions by lingering
    beyond his appointed hour, is now, by all reasonable standards
    used to determine such matters, dead.[1]

2. During those nights holding vigil at my father's hospital bed-
    side, I stayed awake counting his breaths. Human breath, or so
    Aristotle conjectured, supplies the air—the most divine of the
    elements—needed to form the *pneuma zôtikon*, the living spirit.[2]
    I counted as my father's living spirit—still hitched to his wracked
    body—was sustained by twelve breaths a minute; seven hun-
    dred and twenty ragged breaths an hour. When he breathed a
    final time—the last of six hundred million by my calculation—I
    was away. My mother and two siblings were at his side.

3. Years before this, a stroke had left my father alive but greatly diminished. At the Village Coffee Shop, my mother and I drank our cappuccinos and ate our Bakewell tarts, and I said, "It may not seem so yet, but this family has already fallen apart." Later, Dad would tell me stories, each one polished smooth as a pebble caught in the endless tide. Everything unnecessary is rubbed away. Often it is the same story told in the same way: the waves don't tire of the same old moon, the same path along the shore.

4. When I returned to my father's bedside moments after he'd been pronounced dead, I held his hand in mine; it was still warm. An editorial published in the *British Medical Journal* in 1895 responding to a spate of anxiety about premature interment vaguely crossed my mind.[3] Yes, the editors conceded, there *are* occasions when syncope, coma, convulsion, hysterical spasm, catalepsy, and exhaustion mimic the death throes. Assuringly, however, medical professionals could *always* tell when life has truly departed the body. Holding my father's hand, I could not recall what the true signs of death were.

5. My father was not a man for grasping hands. His expressions of affection took other forms. One morning, after a night of childhood terrors drove me to my parents' room, I rose to leave the comforts of their bed when my father stirred. His sleeping hand fell across me. I studied it as if it were a stunning natural marvel, which, I suppose, every father's hand is. As he lay in his casket, I took his hand for the last time. It was so cold. I'd never before known how cold is the lifeless air.

6. What first we learn about fire, philosopher Gaston Bachelard reminds us, is not to touch it.[4] Fire, however, was the last element to engage my father's remains. So hot is a cremation chamber that a body is mostly vaporized rather than burned. Although my father's milled bones were returned to us, I prefer

to think of him in that elemental stream released from Mount Jerome Crematorium, spirited over Templeogue, and carried across the land. Perhaps, a measure of him made it home to Tralee to live on there, bound to the earth of the Maharees.

What Is Life; What Is Death?

7. Impossible to embrace life without gathering an armful of death. I mean this conceptually, definitionally.[5] If to define something puts boundaries around that thing, and says "this far but no farther," then it is clear that life and nonlife are engaged in constant border skirmishes. If fire, for example, isn't alive, then why do we (in English) insist on "feeding" it?[6] If Earth isn't alive, why call her "mother"? That life at the margins of its definition bleeds into the never-alive should not unnerve us; blurred boundaries also circumscribe philosophy, culture, art, even death itself.[7]

8. Physicist Erwin Schrödinger's short book *What Is Life?* is based on lectures he gave in Dublin in 1943.[8] In it he claimed that life's essential property is its thermodynamic improbability. We should question how life "evades the decay to equilibrium."[9] Why don't we instantly fall apart? After all, to be consistent with the second thermodynamic law, *all* things are fated to decay. Schrödinger's answer to this question is profound and straightforward: organisms maintain a high level of orderliness—low entropy—by continually "sucking orderliness" from their environment. The living and the nonliving are indissolubly linked.

9. On Earth as in the heavens, all things reside in the kingdom of decay—it's the ineluctable message of thermodynamics. When you chance upon a thing not falling apart, you *must* ask, "Why not? How are you resisting decay?" When naturalist John

Muir wrote that everything is "hitched to everything else in the Universe," he might have added the less jolly codicil that any well-arrayed thing is twinned to disarray and that disarray is the sibling that pulls the wagon, that constructs the road, and that sets the nose inexorably in the direction of death.[10]

10. Despite our vaunted bipedality, humans move through the world as if on a horizontal plane, nibbling at the world as we go. To the rear, the vermiform tube of our body emits its rivulet of waste. As with humans, so too with other organisms—the stomatal and the mouthless alike: absorbing order, depositing streams of havoc. In saying that life and environment are indissolubly linked, it's the equilibration of order and chaos that is meant. Human uniqueness stems not from our being disruptive; rather, it's the pace and scale of our action that is distinctive.

11. Before James Lovelock speculated about the nature of a homeostatic, life-sustaining, and planetary-scaled entity, that is, about Gaia, he investigated the physical basis of life-detection tests.[11] He conjectured that the presence of life on any planet can be confirmed by the "omnipresence of intense orderliness... and of events utterly improbable on a basis of thermodynamic equilibrium." A source for this suggestion: Erwin Schrödinger's *What Is Life?* If Lovelock was correct, then Earth is maintained by the exhalations of all life combined. Earth is ordered on a grander scale than was previously scientifically imagined.[12]

12. A planet is dead when a steady-state equilibrium is reached for all its chemical potential. On the intimate scale, an organism dies when it ceases to organize its environment and is *itself* reorganized by its environment. To put the matter differently, an organism or a planet dies when it endures a surfeit of decay over production. Death, like life, is not a matter just for the

isolated individual; rather, death is a cosmic affair. For all of that, the moment of death remains both misery and mystery; it is when "the ghost leaves the machine."[13]

Love and Strife

13. The pre-Socratic philosophers Leucippus (5th c. BCE) and Democritus (mid-5th–4th c. BCE) conjectured the existence of immutable primary substances—uncuttable elements: the *atoms*—moving through the void. Changes in plural forms—the compounded multiplicity of nature—stem from the aggregation (growth) of atomic structures and their subsequent dissolution (decay). Later, Empedocles (c. 495–c. 435 BCE) described nature as emerging from opposing processes: "harmonization" into unity and "dissolution" into plurality. The four elements—earth, air, fire, and water—are harmonized by love. Strife is the force of dissolution. One becomes many: things fall apart.

14. "To every thing there is a season," Ecclesiastes reminds us, "and a time to every purpose under heaven...a time to break down, and a time to build up."[14] This intuition of a harmony between creation and destruction is echoed by Empedocles, who declares that concerning nature a twofold story is told: "At one time, they grew to be one alone out of many, at another again they grew apart to be many out of one."[15] Look for "love" and "strife" under their many names; in the natural order of things, they should achieve a balance.

15. The degree to which, in our modern conception of the cosmos, the forces of order and disorder are balanced is a question best left to physicists. Undoubtedly, we live in a universe where the second law of thermodynamics is inviolate, where the arrow of time is drawn from the quiver of dissolution. On the local

scales where you and I operate, we can see creation and destruction play out in fateful ways. We're born, we endure a bit, and then we die. And as we endure, we atrophy, we resist, we repair, only to succumb.

16. The duality, in our daily lives, of the forces of both aggregation and disaggregation—of love and of strife—are so far-reaching that these tendencies are called by many names. Synonyms for the former include *growth, building, constructing, creating, developing, enhancing, expanding, flourishing, improving, proliferating, strengthening, and thriving.* Synonyms for the latter include *decay, atrophy, corrosion, decomposition, degeneration, destruction, deterioration, disintegration, erosion, moldering, rotting, simplification,* and *unraveling.* Look around you. For each object that comes into view, ask which of these phenomena is in ascendency. What is the state of affairs in your immediate surroundings?

17. A conceptual struggle that is foundational to ancient philosophy remains a contemporary preoccupation: how to reconcile the unity of growth and the plurality ensuing from decay. Does an equity prevail in our intellectual interests concerning these poles? A possible means of quantifying the balance is to trawl the literature of the natural sciences for evidence. Thus, in the titles of scientific papers published from 1975 to the present, I searched for the occurrence of sixteen synonyms of both *unity-growth* and *plurality-decay.*[16] The stark results: 348,733 papers relate to unity-growth and a mere 44,303 on plurality-decay.

18. A skewed attention among natural scientists to growth is understandable: scientific endeavors are largely pragmatic affairs. But does the neglect of decay extend to other intellectual traditions? It's not an easy question to answer: the key synonyms I previously used to explore growth and decay are not especially relevant for philosophy, for example. But do ideas of

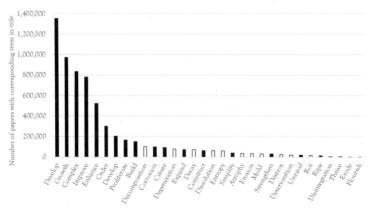

Prevalence of terms related to growth (black bars) or decay (white bars) in titles of research papers in the scientific literature, as found in the Web of Science core collection, 1975–present.

growth and decay not also evoke reflections upon both life and death? Analyzing the publications of biologists reveals that they discuss life twice as much as death. In contrast, philosophers are *four* times more likely to discuss life.[17]

A Return to Earth

19. All that flourishes returns to earth. For a short duration, entities are hoisted up upon the world's thin rind, flourish for their moment, and are promptly resorbed into sediment.[18] It's a lesson I learned young. Although I was not an overly morose child, I inclined in my early attitudes—and later in research endeavors—to decay rather than growth, to the mechanics of decomposition rather than production. Reflecting upon this shadow realm provides, I think, the resources to illuminate one's daily struggles: from mundane matters to those existential disquietudes that keep us awake at night.

20. We shun death; we shun decay. Even intellectuals may shy away. In the modern university, one finds "life sciences"

departments, and even if that faculty fails adequately to capture all life, such departments still aggregate, under the one roof, many of life's quintessential themes. In contrast, death and decay scurry down far-flung corridors. Morbid themes are discussed, fragmentarily, in psychology, environmental science, philosophy, religious studies, social work, and even literary programs. As a result of this intellectual dispersion, when a loved one dies, I've found that the conceptual and spiritual resources were diffuse and inaccessible.

21. For all that, the lessons of decay are all around us: dead bodies, sure, but also leaves, bridges, atomic nuclei, solar systems, cities, friendships, that carton of milk left unattended, the biodiversity of the earth, my aging body, the trellising of my skin, marriages, civilizations both ancient and contemporary, and—driven by the inescapable law of entropic decay—the universe itself hurtles to its heat death when *all that is* comes into final equilibrium; the cosmos, like an old man sinking finally into his twilight bed, reaches, at long last, a constant uniform temperature.

22. There's a moment during dissolution when the properties diagnostic of a decaying entity disappear. That once-dark hair is now unambiguously gray; that once-warm hand is cold. On a grander scale, a stressed Earth may lose resilience.[19] Such properties of entities not easily traced to those of its parts are termed *emergent* ones: the whole asserting itself beyond the essential features of the parts. But as things fall apart—when Empedoclean "strife" prevails—what term might we use when the distinctive properties of the whole depart?[20] Should these nameless departures not also be given their due?

23. "For dust you are, and to dust you shall return"—thus in Genesis God invites us *from the very beginning* to keep the decay of the flesh ever close to mind. The "ghost leaves the machine," the

soul departs the body: the essential property seems to be gone. And yet for all that, in decay something *always* endures. My father's elements were dispersed to the air; some now have fallen to Earth. I remember that polymathic man, I remember his ceaseless stories, I remember a hand fallen across my body. I remember a father's care.

24. Each paragraph in this essay is ninety-five words, one word for each of my father's years on Earth. As I write these final sentences, I am counting down. Sixty-seven words remain, now sixty-four: they slip away almost without my noticing. Only a handful remain. I recollect this: in his final months, as his memories faded, my father would suddenly recall stories from his youth. And not just any stories, but some of the richest, most entertaining stories, ones that had been buried so deep that he thought he'd never find them again. No words remain.[21]

notes

1. RIP.ie, "The Death Has Occurred of Patrick Heneghan," https://rip.ie/death-notice/patrick-heneghan-dublin-templeogue-477614.
2. Aristotle, "De Anima," in *The Basic Works of Aristotle,* ed. Richard McKeon (New York: Modern Library, 2009).
3. "The Signs of Death," *British Medical Journal* 2, no. 1813 (1895): 787–88.
4. Gaston Bachelard, *The Psychoanalysis of Fire* (New York: Beacon Press, 1964).
5. According to the Clarivate's Web of Science, since 1975 the number of scientific papers titled "What Is Life?" is 129.
6. There are other languages in which it seems that "feeding fire" has its counterpart, but I have not done an exhaustive survey.
7. To illustrate this point, one that might not immediately seem intuitive, you might note the number of books and articles entitled "What Is..." (philosophy, culture, art, etc.). In fact, most advocates of major disciplines are embroiled in productive debates about where one discipline ends and another begins.
8. Erwin Schrödinger, *What Is Life? With Mind and Matter and Autobiographical Sketches* (Cambridge: Cambridge University Press, 1992).
9. By this poetic phrase, Schrödinger means to compare the vast orderliness of the living entity with the background chemical equilibrium of the universe beyond the vivid entity, an equilibrium that the organism resists for as long as it lives.
10. The phrase from Muir is given in several places in his writing, including "My First Summer in the Sierra," in *John Muir: Nature Writings* (New York: Library of America, 1997), vol. 2.
11. See James Lovelock, "A Physical Basis for Life Detection Experiments," *Nature* 207, no. 4997 (1965): 568–70. This work contributed to the development of tests for life

deployed by NASA's Viking program.

12. The Gaia hypothesis remains controversial but has persisted and remains influential. The essence of the idea is that each living entity retains its order by exporting disorder, and yet the collective effluxions of organisms exert a homeostatic control on the conditions for life. See James Lovelock, *Gaia: A New Look at Life on Earth* (Oxford: Oxford University Press, 1979).

13. The phrase "ghost in the machine" (used derisively in reference to Cartesian philosophy) comes from Gilbert Ryle, *The Concept of Mind* (Chicago: University of Chicago Press, 1949). Later, Arthur Koestler made it the title of his 1967 book on psychology and complex systems.

14. The translation is from the King James Bible.

15. Empedocles fragment 17.1–2. The edition of fragments from the pre-Socratics that I use is Patricia Curd, ed., *A Presocratics Reader: Selected Fragments and Testimonia* (Indianapolis, IN: Hackett Publishing, 2011).

16. The following search terms were used: *atrophy, corrosion, decay, decomposition, degeneration, destroy, deterioration, disintegration, dissolution, entropy, erode, erosion, mold, rot, simplify, unravel, build, complex, construct, create, develop, enhance, expand, flourish, growth, improve, order, proliferate, ripe, strengthen,* and *thrive.* In addition to the core terms, variants of the words were investigated (e.g., *destroy* and *destruction, construct* and *constructions*). For this search, the Web of Science Core Collection (1975–) was employed.

17. I searched for *life* and *alive,* and *death* and *dead,* in the papers available on the digital repository JSTOR. The prevalence of these terms among biology and environmental journals versus philosophical journals was used to generate the ratios mentioned in this paragraph. In biology (and environmental) journals, *death* or *dead* occurs 322,527 times; *alive* or *life* occurred 611,905 times. In philosophy, *alive* or *life* occurs in the titles of 3,487 papers, whereas *dead* or *death* occurred 913 times. The large difference in the absolute numbers reflect a substantial difference in the numbers of disciplinarily specific journals maintained by JSTOR.

18. Energetically, the first five centimeters of soil are the most active components of an ecosystem. See John Aber and Jerry M. Melillo, *Terrestrial Ecosystems* (Philadelphia: Saunders College Publishing, 1991).

19. What I have in mind in using the term *resilience* is primarily its use in the ecological literature, that is, the capacity of Earth as a system to absorb disturbance without a significant departure from its customary structure and functioning.

20. I call those properties that will depart a complex entity during decay—the ones that had formerly been diagnostic ones—Adeiatic properties (from the Greek άδεια).

21. In these paragraphs, I used Microsoft Word's total for each paragraph (excluding associated endnotes and paragraph number).

The Scat of It

Nickole Brown

The shit of it, the slick of it, the beetle's tumbling joy,
the bear's berry slush of it, the coyote's ghost white
dry of it—undigested fur, nothing more, hot-pressed into a
turd—that *nothing-wasted* prayer. The shame

of it, even the dog shy, peering from behind a bush,
spine curved into the *not-in-my-yard*-sign; the teasing too,
me laughing about *the anal express*, the poor cat hissing
at the vet's gloved hand. The dump and log

slop of it, a sad jaundiced yellow or something rich, a deposit
of iron, green nearly black, the color of a forest
never once cut, miraculously untouched.

Then too there's the zoo—regular factories
of it: the chimp's sling of it against his bars, and not too far
from him, swaying ceaselessly from side to side—the elephant—
how hers is shoveled up, scraped from the concrete floor
then hosed down, the rest of the heft hauled away. Down
    the road

it's sweet meat for the pumpkin patch and hungry rows
of corn. And further on, in the dark of the barn, the halo of it
glows white around a chicken's diddle warming next to her eggs.
The hen broods in, pays no mind to the much more tidy loo

kept by those few lucky pigs allowed to stand
and walk away from their bed to defecate outside, so different
from the *lift-your-tail-and-go-where-you-stand* kind—
that of the goat and sheep and rabbit—each pellet perfectly round,
a pile of dinky moons eclipsed, a mess of shining beads,
    a black rosary

undone, the prey animal take on it—look both ways and shit
quick, no dallying around.

The rice-sized mouse of it in the kitchen drawer, even smaller
is that of the roach, the cabinet scrubbed raw because
    mama says
such leavings are degrading, meant for the dirty and poor.

The satisfaction of it—the full-belly, the *I-did-my-job-*

now-let-go, as in what the earth has given my cells have loved
to death and now give back what's left, a cramp of *thank you,*
here is my offering, a stench maybe for us but for everything else

a bouquet of gratitude, a scattering that if you look close you can
track, at least until it's finally buried again, whipping with
worms, churned in, folded back. There is no shame

in it, and if we are disgusted, we have not yet
learned—blessed is that from what we came, blessed
to what we return.

We Walk This Earth: Land-Based Indigenous Education Practices

Tia M. Pocknett

S itting outside, drinking from my water bottle, I look out to my students who are also enjoying a sip of water. We have hiked a mile and a half thus far, and they are looking a little tired. My students, who are in the age range of six to ten, are wide eyed: screening the woods, looking at the ground, they are quiet, inquisitive. I am proud of them. I remind students to keep to the path as we take a break, telling them that if we go off the path, we are more likely to pick up some unwanted hitchhikers, such as ticks.

Many of my students understand how we walk in the woods, making sure not to stir all of creation but to walk in harmony with creation—to be careful with their footing. Many of these students have hiked with me before, especially my older students, who have been with me since 2016, almost seven years. These students have been in the woods, on nature walks and hikes with me since they were two and a half years old. They can identify all sorts of plants and animals we see on our hikes, even understanding medicinal properties of some of the local flora.

Going on hikes is an integral part of our school-year curriculum. Walking on the earth connects the children to the trees, sacred medicines, the animals, and, most importantly, to our ancestors. This experiential learning process is essential to our students. For students to have the understanding that we are all connected, we must make sure we are connecting naturally to the world around us. Walking the earth, having our feet planted into the ground, we feel our mother as we walk, we are reminded of our ancestors who walked this earth before us, and we are continuing to keep the trails for the next seven generations. Personally, I call upon my upbringing and the ones who taught me. I remember moving across the land and water with community members, some who have crossed over into the spirit world. The teachings that were provided to me I know are important and must be passed on. That is the work that I want to do; that is what inspires me to work with my community.

Weetumuw School: Montessori Pedagogy Fused with Wampanoag Culture

These students are a part of the Weetumuw School, a Mashpee Wampanoag language and cultural immersion school that teaches children age three to nine using a Montessori pedagogical framework—all while keeping true to our traditional teachings.[1] My colleagues and I have developed cultural and language curricula that aligns to sacred teachings from our ancestors as well as from the living world around us. The school is divided into two classrooms: Mukayuhsak Weekuw (Primary House, for ages three to six) and Wushkeenune8ak Weekuw (Lower Elementary House, for ages six to nine). My classroom is Wushkeenune8ak Weekuw.

Wushkeenune8ak Weekuw was established more recently than Mukayuhsak Weekuw. My colleague Siobhan Brown and I opened Mukayuhsak Weekuw back in the fall of 2016. Because of the many requests from parents, Siobhan opened Wushkeenune8ak Weekuw in 2018 and then passed me the mantle of the classroom in the fall of 2021. This year, I am enrolled in an upper elementary Montessori training program, and after the training is complete we hope to open an upper elementary classroom for children ages nine to twelve. Opening a new classroom requires us to begin exploring how we will tie our cultural teachings to advanced western teachings that students might experience in a public school. We must develop relationships with members of the community who are working on conservation and consult with those practicing sovereignty rights, and those members who are working towards reclaiming and acquiring lands to return to our nation, which will help guide our cultural teachings for upper elementary children. Most importantly, our elders hold the wealth of knowledge in regard to our land and our traditional practices. Creating a tribal community school requires the expertise of our nation to draw from important, traditional knowledges and pass them along to future generations for years to come. This requires us to build with our community and just for our community.

In 2021, working with members of the Mashpee Wampanoag Community and with fellow colleagues, we developed a cultural framework for both classrooms. The overarching theme for the cultural framework is working with natural seasons in our geographical area. The Gregorian calendar is not a natural measure of time for Indigenous people, plants, animals, fungi, the waters, and Earth; our ancestors were forced to use the calendar as a way to assimilate with our colonizers, and it ignores our traditional teachings, ways of being, and understanding of the natural world around us. Wampanoag people always looked toward these seasons before the Gregorian calendar had been developed. Evidence of their practices are stones that we go to that align with the sun and tell us when to plant, when to harvest, and when to move back into our winter homes.

As we continue to develop our cultural framework, we are looking at the vast teachings that come from our outdoor world and from our culture keepers. Many traditional culture keepers are finding that these stones, even though they do align with the sun, have changed so much due to climate shifts that sometimes relying on them isn't enough. Looking at natural signs from plants shows that in our area the planting season starts a little earlier. Geographically, it is much warmer here in the summer, and our winters are no longer as cold as they once were. When I was younger, I remember always having snow on the ground. But now, in winter, we are lucky if it snows once or twice within the season. This affects all plants and animals, which also affects us as well.

Creation Stories and the Formation of Land and Water Ways

When I was a little girl, I remember hearing creation stories that described how the land was created. One of our traditional characters from our story was Maushop:

> Maushop was a giant who traveled all across Turtle Island. He dragged his foot with sand out into the sea, forming what we now know as Cape Cod. When he was finished, Maushop sat upon the beach, tired. He took off one moccasin and poured out the sand from it into the ocean. The sand became the island of Martha's Vineyard. He then took off his other moccasin and tossed it into the sea as if to say he was done walking. When his moccasin landed in the water, it became the island of Nantucket. Maushop grabbed his pipe and then smoked it, giving thanks to creator. Now, whenever we see fog at Nantucket Sound, we are reminded of the formation of the land, and we know Maushop is smoking his pipe, in ceremony giving thanks to Creator.

This is the story that we continue to pass down to each generation. Children look at the land formations on a map and can point out the island that looks like his moccasin, Nantucket, and the other island, which was from the sand he dug out to form Cape Cod, Martha's Vineyard. When the weather is warm, we bring the children to the beach so they can see the island at low tide, as it is the most visible then, and consider whether this was the beach where Maushop sat. It is of great importance to bring the children to these areas so they can begin to understand a deeper, spiritual connection to the land, can maybe envision the land vacant. Having the students try to visualize the land without these homes breaks down these structures of colonization and calls upon the students to think about a scenario of "what if?" What if we weren't colonized? What if we still lived in our traditional homes? What if we still spoke our language? What if we moved upon the land and waterways like our ancestors?

Another story I love to share with students is the formation of the lake Scargo, which can be found in Dennis, Massachusetts:

> Long time ago, there lived a beautiful girl who made friends with a fish from her community pond. Each day, she would go to the pond and swim with her fish. Soon she noticed that the water level of the pond was getting lower and the fishermen from her community were unlucky catching fish to feed their community. However, the fish she befriended was growing abnormally in size. After many days with no fish to feed their community, people began to die, but the fish the girl swam with still grew in size and the young girl fell in love with the fish. Each day, as the water level diminished, the fish would grow, and more people would die from the young girl's community. The young girl became sad and told her fish about what was happening to her family and community. Not wanting to see the young girl sad, the fish told her of a sacrifice he was willing to make. For he loved the girl as much as the girl loved him and he too was afraid to see her perish from the Earth. He told her that her community would need to feast from him. Heartbroken, she argued with the fish, but, in the end, she followed his words. She told the fisher people of her community, and they harvested the fish from the water.
>
> After the feast, her community was fed but the girl was still sad. She knelt by the pond, whose waters were extremely low, and began to weep. Her tears filled the pond and reshaped its boundaries into the shape of a fish. As her tears filled the pond, fish returned to the pond and the girl's community never went hungry again.

When the students visit Scargo Lake, we bring tobacco offerings, and we retell the story to the students. There is a tower close by, so the students are able to climb the tower and look over the lake. From that vantage point, students are able to see the shape of the water, and the stories become true. If we only tell a story and do not actually go to the places we speak of, then it just becomes an ordinary legend. These stories and traveling to local sites validate our traditional teachings.

Throughout the school year, we travel all over Cape Cod and all the southeastern parts of Massachusetts, like our ancestors, visiting sacred rocks that have been here longer than any human, as well as beautiful cedar swamp forests and pine barrens. When we make these visits, we retell countless stories our family members have passed down to us. Our hope is that the students will continue to pass the stories down and continue to occupy these sacred lands for generations to come. Telling the stories is not enough; bringing the children to these areas and allowing them to explore and connect is the true work.

Applying Conservation Methodologies to Our Outdoor Curriculum

When we visit these sacred locations, we are also looking to see if the earth has been taken care of. We always bring a trash bag with us, just in case we see litter. Our students understand that trash is not natural, and that our ancestors were big on recycling and reusing our resources. When the students see the land littered with trash, they are confused. They do not understand why others do not care for the land. Many times, I will be asked, "Sânushq Tia, isn't our job as human beings is to take care of the rest of the beings?" or "Sânushq Tia, why do people not care about our earth?" I remind the students that not everyone receives our sacred teachings, some learn too late, and others may never learn that, to be in harmony with the world around us, everyone must work together to take care of the place we live in.

Sometimes, visiting different places all over Cape Cod, we have to pick up small bits of trash. Other times, the damage done to our land isn't something twenty-four small hands can fix. Recently, we have been visiting our waterways, which have become polluted by lawn treatment runoff and septic-system waste that is pumped into ponds, bays, and rivers, specifically our Mashpee River. During a recent walk over to Punkhorn Point (where Popponesset Bay feeds into the Mashpee River) to visit with our tribal Natural Resource Department, it was pointed out to the students how the pollution is affecting not only what lives in the waterways but also the earth under the water.

Black sludge created from the algal blooms is affecting not only the waterways but also the plants by the waterways, which then has an effect on the animals that live off the plants or look to them for survival. The *New York Times* writer Christopher Flavelle was able to document and report on the water quality that inadvertently affected the earth around the water.[2] The poison that is coursing through the river affects not only the fish and shellfish we eat but also a wide variety of plants and animals who are unable to sustain the life that depends on them. Natural Resource Department workers recall a time when shellfish and eelgrass were plenty. Now the eelgrass cannot survive the muck that settles on the bottom of our local waterways, and the once-abundant shellfish population has been devastated.

So, what's the lesson the children learn from this experience? We are able to see these polluted areas while trying to forage for plants and shellfish. Learning that we are unable to harvest shifts the lessons from food sustainability to conservation tactics. In partnership with the tribe's oyster farm, the children are learning that seeding the river with oysters helps reduce nitrogen levels, which will in turn bring down the algal blooms. The children want to take care of the river, the shellfish, the fish, and all the plants that run with the river. They feel that it is necessary work for their families to be healthy so that they are healthy and so that one day, when they have families of their own, they can return to the same river and eat as our ancestors did.

Benefits of Land-Based Education

We are finding that our students enjoy their time at school when we are outdoors. During our week, we are always outdoors: recording weather, studying trees, going on hikes, or simply reading outdoors. There are times when we find ourselves enjoying exploring swamps, marshes, pine forests, beaches, and other landscapes, and there are other times when we become sad when we find the earth sick. Cultivating a sense of pride and the call to take care of the earth, the children are actively finding ways each day to protect the earth and all of its inhabitants. While out and about, we bring along a trash bag and return to school to sort the trash between what can be recycled and what cannot.

Recently, we found ourselves at a beach in South Mashpee. Sometimes, I like to take the children to the beach, so I can catch my breath as an educator and calm down in the ambience of the beach environment. Listening to the waves crash against the shore, with the warm sand underneath me and encasing my feet, the wind subtly blowing against my face, and the warmth of the sun is the peace I am always searching for. It is here on this planet where I feel the most connected to the earth. However, as a teacher, that peace never lasts long. As soon as we get onto the beach, I take my shoes off to get my feet into the sand and all of a sudden I feel the ambush from my students as the questions begin to be hurled at me. "Sânushq Tia, what is this?" "Sânushq Tia, did this beach become eroded by water, wind, or water and wind?" "Sânushq Tia, what kind of shells are these?" "Sânushq Tia, can we eat these?" "Sânushq Tia, what did our ancestors do here?"

I am happy to answer their questions and grateful that we can talk about the land and water in the outdoor environment. Ideas about what our ancestors would eat, and what our ancestors did on the land, no longer feel abstract as we begin to teach and demonstrate exactly what they do on the land.

For example, the other day I brought my students to the Nauset Marsh Trail located in Eastham, Massachusetts. As soon as we got on the path, we were met by the marsh. Students were able to tell, without the help of an adult, whether it was high or low tide by their lived experiences from a sensorial point of view—the smell of the marsh; the view of the land, which included some shellfish we could see; fish that didn't catch the slack of the tide and were caught on the land during low tide. These are all physical evidence that students point out to one another as they deliberate whether it is high or low tide. Once the students have a consensus on the state of the tide, someone will pipe up with what they have decided; 80 percent of the time they are right.

As we continued onward during our hike, the landscape changed and the children were quick to point out the physical land differences that signal environmental change. We pointed out the trees, plants, and animals that are the same and different in both areas. When walking closer to the marsh, we can see the seagulls and the many grasses that grow in the marsh. While walking inland, students notice the trees begin to change—pointing out how there are more cedar trees that are closer to the marsh—and as we walk further inland, the landscape of trees begins to change from cedar to pine, oak, and maple trees. Students notice that they no longer hear the gulls cry but hear song birds. On one of our trips, my colleague pointed out the types of trees our ancestors use for building Weety8ômash.[3] He talked about how the trees are processed. Because it was spring time, it was the perfect time to peel cedar bark. We put down some tobacco and thanked the trees.

My colleague took his knife and demonstrated how to peel the bark. Once we had peeled the bark from the tree, we demonstrated to the children how to process the bark, showing how to separate the outside of the bark from the inside. We made coils with the inside of the bark, saving them to later show the students how to boil them and then weave with those materials. The outer bark we would use to teach how we dye cord for weaving.

Children get to experience the process while being as present as possible during these teachings—a sensory learning process. These teachings are lessons not only in environmental science but also in botany, chemistry, fine art, history, and anthropology. The process through which we educate children on our land isn't just a simple walk, it is a whole-body, multicurriculum experience that integrates traditional Wampanoag and mainstream Montessori teachings.

When these tiny hands process the bark, it is also a work of healing from generational trauma. A lot of our teachings were stripped from ancestors through forced assimilation. Wampanoag people have dealt with colonization for over four hundred years. When our children learn these practices and teachings, they take them home and share with their families. Sometimes, when the children share with their extended family members, those family members will recall moments from their childhood in which they remember gathering materials, foraging for medicinal and edible plants, ways they fished and hunted on the land, and how they built sacred spaces and practiced sacred ceremonies with family members who have passed on into the spirit world. Sometimes, from what the children have experienced at school and brought home to their families, we find that the parents want to connect with these teachings as well. Most importantly, this work folds the family back into the tribal community and helps us maintain healthy relationships with our children, our extended families, and with all creation.

This cultural reclamation work is just as important as language reclamation work. When we as tribal people are reflecting on what we need, people talk about teaching children our ways so that they know their history and that they continue the work of our ancestors. Language reclamation tends to be better funded than cultural reclamation work. In the United States, we haven't truly begun reconciliation work with tribal Nations. Only when the federal government can begin to acknowledge and address the atrocities that have occurred to all indigenous people of this land can the work of cultural reclamation truly begin. The federal government would have to admit that forced assimilation through processes of genocide, relocation, boarding schools, and blood quantum contribute to language loss and most importantly to our sacred teachings being stripped from all indigenous people. Having these teachings thrown aside by our colonizers and forcing western philosophies and religious teachings upon us has resulted in both the loss of language and traditional oral teachings from our ancestors. Language reclamation and cultural reclamation work is something that I personally view as work that goes hand in hand. Language reclamation work allows us to dive deeper into understanding how our ancestors viewed the world, the land, the trees, and the water. Both language and cultural reclamation work help us reclaim a holistic traditional educational experience.

Being in conversation with elders, community members, and other indigenous people has inspired me to want to work with my community, especially with the children. As long as I walk this earth, I will continue to walk the earth with the future generations, teaching them all I know to ensure our community's vision is being passed on and to make our ancestors proud knowing their struggles meant something.

Caring for the earth is a practice that everyone in the world should be tasked with. We have only one planet, and it is ours to take care of, to ensure that our children get to experience a healthy earth. Teaching children the importance of clean water, clean land, land conservation, and climate change and how it truly affects all people is extremely important not only for indigenous people but for all people, as well as all living and nonliving beings on this planet. When students experience their feet on the earth, their hands on trees, their eyes on the beauty of the plants and wildlife around them, they open up possibilities that future generations will work toward keeping our planet clean, protected, and loved.

notes

1. The Weetumuw School is a Mashpee Wampanoag tribal culture and language immersion school located in Massachusetts.
2. Christopher Flavelle, "A Toxic Stew on Cape Cod: Human Waste and Warming Water," *New York Times*, January 1, 2023, https://www.nytimes.com/2023/01/01/climate/cape-cod-algae-septic.html
3. *Weety8ômash* is the Wôpanâak word for "houses," also known as *wetus* or wigwams.

Luminous Ground

Andreas Weber

G rowing up, I spent my best hours with my hands in the soil. I leased a vegetable garden at the age of twelve. Before I became caretaker of that plot, a neighborhood family farmed the triangular piece of land between the road, a wheat field, and the parking lot. I often looked over the fence when I saw them working in the garden, wishing I could do something similar to what they were doing, although I didn't know exactly what cultivating a garden actually involved.

My parents had planted bushes, shrubs, and a few small trees behind their terrace and, above all, had grown a lawn, like all the other terrace-house owners. The only kind of gardening I knew consisted of pushing a rattling hand mower over the grassy estate, for fifty pennies of pocket money per deployment. Nonetheless, I was sure that working in the soil and among the growing green plant bodies had to result in something delicious.

When the neighbors unexpectedly moved away, the garden fell fallow. The soil quickly was overtaken by weeds. Three rows of strawberries disappeared under tall couch grass. I could see their fruit shimmering between the stalks as summer arrived. Maybe that was the trigger for me to take action. I convinced my father to lease the garden for me. The rent seemed ridiculously small, even to me as a child, for what I received in return. I had a garden all to myself.

I started right away. That is, I obtained literature. The *Self-Sufficient Gardener* by the British author John Seymour became my catechism. But I bought more. At the apex of my gardening enthusiasm, I owned an entire shelf of literature on vegetable gardening. If permaculture had already been popular back then—in the early 1980s—I would have delved into its theories with rapture and eagerly put its principles into practice.

The books were not a substitute for physical labor, or an excuse to postpone it. Rather, reading intensified my enthusiasm for working in the garden. I could turn twenty square meters of ground during the day and flip through dozens of pages before I went to sleep at night, checking how John Seymour would do it. In bed with my books, I regularly concocted topographical schemes. I drew planting layouts, dreamed up ambitious innovations, distributed the available land mentally over and over again to carrots and kohlrabi, beans and borage. Thinking about these plans doubled my desire to touch the ground with my hands the next day.

The earth took care of me. In the first spring after I had gotten access to the land, I started molding the wet soil as soon as the frost had subsided. My father briefly showed me how he had learned to turn the earth as a child, and then he left me alone after that. He didn't come by much, and when he did, it was usually to point out something I hadn't done the way he thought it should be done. The chain-link fence was too low. The bed borders bothered him. But I didn't listen. I had begun to receive a different form of parenting.

On my little patch of earth, I was free. I could try everything without adult oversight or input. And I had to—I needed to start from scratch. In winter, I dreamed. When spring approached, I went to the garden store, with its very specific scent of humus, plant sap, fertilizer, and poison, and bought a few bags of seeds: lettuce, radish, parsley, dill, leek, cucumber.

To this day, I remember the feeling of fresh soil on the skin of my fingers. I had tilled and meticulously raked it so that it broke into marble-sized crumbs and dried slightly on the surface. Bereft of its moisture, the soil took on a muted gray. Freshly broken open, it was jet black. Today I know that this land was extraordinarily fertile. The earth said so through shimmering moisture, fragrance, and the numbers of fat, red earthworms that writhed between the clumps when I dug into the ground.

I spent long afternoons kneeling among the beds, pulling grass and herbs out of the soil. The ground crumbled in the places the roots formerly occupied, opening into small craters. The earth covered the skin of my hands, and it tangled in the fabric of my clothes, as if the soil, too, were a being that somehow communicated with me and that changed me as I acted on it. At the time, I didn't think too much about the distributed, grainy creature that pulled at my skin and held my attention. With childlike pleasure in logistics, I dreamed of new tools (a spading fork!) and exotic vegetables (tomatoes always had to ripen on the kitchen windowsill in those last Holocene summers in northern Germany).

And I was not alone. I had human and more-than-human companions who shared my passion. My best childhood friend often helped me with the worst chores. He was a much more sturdy worker than I. And the family dog insisted on being part of Team Earth, too. Hardly had I dragged the garden tools from the cellar to the front door, when my dachshund stood waiting at the entrance, already tail wagging. While I stirred the crumbs, the dog devoted himself to his own kind of earthwork: he dug after mice. In fact, the dachshund saw our activity as a kind of communal labor. For him, it was important that I did my part. When I paused at the spade to wipe the sweat from my brow, he yelped at me. Only when I plunged the blade into the earth again would the dog contentedly return to his hole and continue to dig further.

A blackbird was also involved in our gardening. He sat on a post of my too-low chain-link fence, waiting for me to unearth a worm. He also scolded me when I slowed my pace and did not produce fresh food quickly. So, I had four work companions, and only one was human: my best friend; then the dachshund, who, from his point of view, was united with me in a massive project of digging through the land for its content of mice; the blackbird, for whom this project meant the search for worms; and the earth, which carried, nourished, and connected us all. At the same time, the earth constantly undermined our efforts, because the soil alone decided what kind of life it was going to produce and support (knotweed, nightshade, goutweed). But in truth, we all, animals and plants, were its various modes of appearance. We all were this Earth.

In the few years during which I tended to my little garden as a child, the earth helped me grow up. It allowed me to partake in its aliveness, simply by allowing me to care for it, to help plants grow in it, to take it into my body in the form of vegetables, and to share its fertile allure with my companions. I learned a lot then. But I did not fathom how much I received. Only now, through the eyes of an adult, I see how my experience as a young gardener have taught me about life and my place in it.

The life that comes from the earth is organic and mineral. Both organic and mineral material are fertile, both depend on each other, both exist through each other. Both follow one desire. Earth teaches us that to be a plant is nothing else than to be a mineral, that to be flesh is nothing else than to be dust. Not because, as Christian theology claims, our flesh is ultimately dust and only our spirit is worthy and capable of salvation, but because that dust comes from the same origin as all bodies. The earth is ultimately our sensitive flesh, the skin of the world, which experiences itself in the touch, in the flow, in the whirls of matter.

This closeness of humans to the fertile soil is enfolded in the deep earthly roots of the English language. The words for topsoil and for human, *humus* and *Homo*, have a common origin in Latin. The adjective *humanus*, "human," "humane," recalls this proximity. There is also a dimension of soul in earth, mirrored by words: *humoral* means "corresponding to the bodily fluids," soul-tuned by the digestive transformations of the body. *Humor* is about our emotions, the connecting terrain where we meet others. And the adjective *humilis*, "humble," sees us bending down to the ground, giving it our labor and care so that we prove ourselves worthy of receiving something in return.

Likewise, I imagine that the word *culture*, when explained from the perspective of the living soil, can be understood as a "cultivating": a tilling of the soil that consists less in pulling heavy trailing gear over the field than in the reciprocal tender care between hands and humus. The earth not only nourishes us but also can teach us how to relate. The ensemble of the ways in which a given community nurtures relationships among its human and more-than-human members can be called culture. Culture, understood as Earth-being, is mutual tenderness.

The topsoil—what we often most simply call earth—is an intricate meshwork of organic and inorganic layers. This soil is like our own bodies, which contain minerals as well (our bones and teeth, for example, but also the salts in the cellular liquids that makes our nervous system function). On average, most productive agricultural topsoils contain 45 percent inorganic stuff: grains of sand, pebbles, dust. About a quarter of it is made up of larger, smaller, and tiny air bubbles. Another quarter is water. On balance, the humus content of the soil is a good 4 percent. Half a percent is roots, those body parts of plants that inhabit the ground. Only a quarter of 1 percent of the soil volume consists of mobile organisms. Most of these are tiny, but they are immensely numerous and species rich. A single shovelful of soil can contain more species than the entire Amazon rainforest aboveground. The majority of these microbiota, of course, are bacteria. Even a single plateful of soil contains more individuals than the world's human population. The weight of microorganisms under the surface of a soccer field equals that of two cows.

In the soil, and in ourselves, what is useful to one individual and what is useful to another cannot be separated. This earthly togetherness is similar to the relationship we entertain with endosymbiotic bacteria in our intestines. Although these are genetically not part of our body, they nevertheless fulfill essential functions so that our shared life processes can be maintained. They are us and they aren't—but without them we would not be. Our intestine is the earth inside of us. Indeed, the interior of the digestive tract, although topographically inside our bodies, is a tube filled with what is outside of us, curving and bending for several meters inside the human body.

The soil, in turn, is in many ways a digestive organ. The spheres of life overlap and interpenetrate and open up to mutual edibility. What is one's own? What is foreign? The world of the earth, the realm of telluric magic, is not a place of fixed possessions: it is life shared, a commons-creating process that constantly produces itself through mutual transformation. In this, the inside of the earth is not only very similar to the inside of our human bodies; it also has affinities to the inside of the cell. All these are tangles of inside and outside, "me" and other, death and birth, which only together make sense.

Ultimately, the individual participants of life originate from the same essence. We are all stuff, matter, that sediments into the ground and becomes soil again. The earth proves that the organic and the mineral are not separate domains. We can see this by closely examining soil particles. What I as a child experienced as earth crumbs between my dirty fingers, scientists call "microaggregates." And those microaggregates, those tiny clumps of soil, are not dead stuff, but wildly alive. They are dust and skin at the same time, mixed in an inextricable way. In a microaggregate, it is hard to distinguish what part is sand from what its living inhabitants are. Scientists can try only to draw a clear and unequivocal line—but in doing so, they risk destroying the tender crumb and its living activity.

Like cells, the mineral assemblies of the soil consist of a mixture of animate and inanimate particles and are, at the same time, sealed off from and intimately connected to the outside.

Between the tiny specks of sand and clay in such aggregates sit minuscule organic clumps: mixtures of bacteria, of juices excreted by living things, and of minerals. Fungal mycelia and root hairs spin around the clumps and give and take nutrients, just as though the clumps were parts of their own bodies. The earth is mineral *and* it is alive.

The same holds true when we turn our perspective away from the ground and to our own flesh. Here, the cells of our body are very much dust aggregates in miniature: crystalized assortments of minerals, proteins, and water that are so intimately entangled that they form inseparable wholes. Although Biology 101 may suggest cells are water-filled containers, our body liquid does not slosh through cells in droplet form. Instead, each water molecule is spun into the molecular structures of the cell. The solid and the liquid, the organic and the inorganic, cannot be separated from each other—neither in the soil nor in biological substance.

At some point everyone is always connected to everyone else. My flesh is also that of the tomatoes. It consists of the same carbon atoms but in an altered configuration, which allows me to experience a different set of individual sensations. I can feel through it; it is a kind of nervous system, allowing the experience of the other as experience of the self. That means that Earth itself is a nervous system, too. Because I am part of the earth, I sense through it, and, conversely, I am a perceptive organ of the soil. I grasp that to exist as mineral is to be constantly transformed in such a way that one's experience is born as flesh. The pulsating cell is nothing other than a grain of sand which faces itself in transformed form.

The ground is a nourishing matrix. Everything in the earth is encompassed in its matrix, which always brings itself forth anew: a maternal milieu from which things are brought forth, are nourished and nurtured, and a milieu that is endowed with the tender longing to be fertile. The English word *matrix* is originally Latin. There it means "womb" or "mother animal." The term *matter*—from Latin *mater*, "mother"—has the same linguistic root. The womb is the "maternal" in its original form as stuff, as world-body, that which transforms itself out of itself, trusting in the life-affirming magic of this metamorphosis.

In an interesting parallel from Buddhism, the original meaning of the Sanskrit word for "emptiness," *Sunyata*, is exactly that: a hollow that fills with a swelling wholeness, like a womb filling with life. Sunyata describes the nature of reality in Buddhism, which is the nature of the mind, too, also called "the ground." By attaching the notion of vacuity to this core dimension, a much vaster and more accessible meaning has been lost. It has been lost in translation in the same way that the maternal ground under our feet has been buried under its use as a substrate for growing things. The Buddhist conception of the ground is described more accurately as something boundlessly generative, playful, and luminous.[1] And I would argue that this provides an apt description of the earth—the ground we walk on, came from, go to—always bursting forth with mobile shoots, letting "emptiness blossom like a thousand flowers blossom," as the Buddhist sage Dōgen put it.[2]

When I look back at those childhood years that I spent in the little garden, I now understand that the earth itself was parenting me. It taught me how to relate—by gently showing me that relating always means to change, and to allow this change to the other. The earth taught me that to stay in connection, I cannot remain the same. Can the earth also teach us how to create cultures that invite aliveness? What might it mean to adopt a mothering attitude and to emulate the way in which matter, *mater-ia*, unconditionally holds us? The maternal ground is productive, unwilling to exclude, and radically vulnerable. The maternal ground is the embracing attitude of matter, and the embracing attitude of that which is fertile in me, *as* me. Thus, the maternal, as a cultivated disposition and as an embrace of life, is the model for taking care of fertile relationships, and thus the model for culture.

Matter is not lifeless; rather, it swells with generativity. The ground is not splintered into anonymous particles that compete with one another. Rather, it is a nimble fabric in which everything merges into the other and everything continues to transform itself through the other. As a child, when I held small chunks of garden soil in my hands, in some sense I also held the elemental Earth—and, in a way, it held me.

The earth still holds me. That's what I think today. When I was a young boy in my vegetable plot, I didn't know how to articulate this. I was simply weaving myself into the soil's fabric, as was the dachshund who yelped at me when I paused our work, as was the blackbird with his radiant eye, dancing up and down until he found a worm—boundlessly *here*, a lithe mold of jet-black earth, nothing but luminous, empty cognizance.

notes

1. I have explored this in depth in Andreas Weber, "Beyond Emptiness: 'Compassion' as the Hidden Ground of Francisco Varela's Thinking," *Journal of Consciousness Studies* 30, nos. 11–12 (2023).
2. Quoted in Llewellyn Vaughan-Lee, *For Love of the Real: A Story of Life's Mystical Secret* (Inverness, CA: Golden Sufi Center, 2015), 48.

And the Ground Opens
Its Mouth to Speak

Jessica Jacobs

"[And God said to Cain,] 'Cursed are you from the ground that
opened its mouth to take your brother's blood from your hand.'"
   —Genesis 4:10–11

Dear wandering dust, dear vagrant clay,
dear humans made of me,

how quickly you've forgotten.
I am not just a backdrop
for your horrors—

read your holy book: Stars and trees
join in battle, hills mourn, valleys
and waves shudder and writhe

at the approach of God. And how
many of your slaughtered
have I choked down?

Your clearcuts evict owls,
salamanders, wolves so you can build
your houses in hills now primed

for fire. I am trying
to warn you. For every season,
I send wrong weather, drain

reefs of their color, let whole species
go extinct. Yet you go on.
Enough. Too much.

Protagonist, delinquent. Who are you
in this story:

Seeing something he wanted
across the road, a boy dropped
his mother's hand

and ran into the snarl
of traffic. She screamed his name,
rooted there, unable to look away.

At the clamor and rush, at a mirror hissing
so close past his ear it raised
the small hairs inside it,

he ran back to her. Weeping,
she slapped him, hard; weeping,
he pressed the heat of his cheek

to her chest. That slap? Pain now
to stave off worse later. A mark
to carry with him and remember.

I am so tired of being afraid
for you.

Becoming Earth:
Experimental Theology

Robin Wall Kimmerer

*J*ust before my turn to the research forest, I pass a little church be- *neath a circle of spruce, with a line of cars in a parking lot that is more dirt than gravel. Out front, block letters on the white wor- ship sign spell out*

Blue River Gospel Church

"In heaven we are released from death"

Bible Study 7 pm

Around the coffee urn inside, I imagine Pastor Ford and a dozen or so parishioners are deep in conversation. "You see," he says "none of us can know for sure what heaven will be like, but we are promised that in heaven there is no time. When death is conquered, time just stops and turns to life everlasting." Old Edna Crosby raises her hand next. "Pastor, when we rise to heaven, what body do you suppose we will be in. This old thing here, or maybe whatever age we want?" The others smile. "Would Roger still recognize me?" Pastor Ford takes her veined old hand between his own. "Oh, Roger will know you all right. When we leave our earthly body, we are transformed."

Just down the highway, at the Andrews Experimental Forest, a cadre of scientists are wondering the same thing.

Trees fall. We know that. But we don't expect to be there, to stand beneath with open mouth to see the lean begin. Down it comes like a slow-motion hammer stroke. In a rush of swooshing branches, the tunnel of darkness closes in and the roar settles to silence. Such peace. I have stumbled into someone else's heaven: the afterlife of cedars, the paradise of firs.

So, this is what it's like to pass through that door, across the threshold of being. The pearly gates are framed by moss-draped firs. I am surrounded by the pillars of ancient trunks backlit by overwhelming radiance. I listen to celestial harps and hear thrushes singing and the music of a distant brook. I see no angels but the moss is a pillow of green cloud.

If I gave it any thought at all, I assumed there was a bright line between the living and the dead, a boundary we cross but twice, once on our way into life and once on our way out. But here the line is blurred. In the afterlife of cedars, nothing is ever dead.

Raindrops glisten from the branch tips and like tiny prisms scatter rainbows through the air. I walk through gardens of forms unknown to flowers, the flora of the dead. Fingers of golden yellow, wrinkled blue toadstools, and fairy rings of translucent gray, a host of flared goblets the color of pumpkins, spiky coral fungi, dabs of jelly fungus, tawny boletes, fluted white bowls of chanterelles, and tiny red cups splattered like blood drops on the moss.

All around me lie waist-high logs draped by blankets of moss. They lie in neat cemetery rows, biers uniform in shape and size. It is as if a great hand had laid them gently on the beds of moss and garnished them with clumps of funeral parlor ferns. They are surrounded by their towering kin, looking glumly down at their fallen relatives in repose.

They say that at the threshold, when life veers toward death, that the story of one's life unfolds before one's eyes. The human brain stores memory we do not fully understand, but the memory of trees is clear. The whole history of each tree is written in the rings. Where the trunk has snapped on the fallen fir, the break is clean and sharp. My thumbnail moves from ridge to ridge, counting off the seasons. Here a wide ring when rains were plentiful, and then three more to mark the years of drought. This distorted black wave is a memory of fire, the thin red band—a year of bark beetles. Fires, windstorms, times of plenty, and times of poor are written here, from the wide rings of youth to the slowing growth of old age.

Does the tree hold some memory of the bear that slept beneath it? It does. Not in a visible ring, but in the body of the cells. Every exhalation of that bear, or the chattering squirrel—or the kneeling scientist with a clipboard—released carbon dioxide into the misty air. Which was absorbed by the open stomates of the fir needle, where it became a building block for the cells my fingers run across.

The log itself is breathing. The carbon dioxide emanating from the microbes in the log is almost instantly absorbed by the pellucid leaves of mosses, who are interlaced with fungi that only moments ago exhaled that same carbon from the log. It's a dizzying circle of inhale and exhale, reciprocity between living and dead.

The very air vibrates with its intake and exhale. There is a deeper sound, a certain hum beneath the silence. I've heard that the planet makes a sound, a vibrating chord in C♯ minor. Could it be the hum of life being made and unmade, composed and decomposed? If we listen very hard, can we hear the soaring sunlit chords of photosynthesis, the countermelody of decay? The quiet is so intense, it is as if I can hear the small suck of carbon dioxide entering a leaf, the prick as a fungal strand breaks through the wall of cedar tracheid.

As inert as the logs seem, there is a ferment of activity inside, like dreams moving inside the head of a sleeper. Fungal mycelia follow the shape of neurons and burrowing beetles unearth memory. Its invisible from the outside, but as a visitor to the afterlife of cedars, I can easily see inside.

Fungi are not alone in their feasting upon logs. Bark beetles, grubs, and carpenter ants riddle the log and open it up to others all too eager to colonize. Bacteria, algae, mites, and spiders—whole food webs develop within the weakening log. When the tree was alive, most all of the cells in the trunk of the tree were dead. They were just empty tubes designed to hold up the tree and to transport water. But now that the tree is dead, it is more alive than ever before. Chipmunks, skunks, and slugs live within. Moss drapes the outside like a shaggy green afghan and ferns sprout from the ends. What was once dead tissue is now a garden, on its way to becoming soil.

Through the misty woods a figure approaches, a slender, gray-haired man in khaki pants and work boots with many miles upon them and an auger in his hand. I'm not surprised at all. This is a man you'd meet in heaven.

The felicitously named Dr. Francis D. Hole was a soils professor at the University of Wisconsin. We were not allowed to use the word *dirt* in his class. Filth had nothing in common with soil so clean and nourishing you could eat it.

"Do you feel that?" he asked, bouncing a little on his toes. "That sponginess underfoot? The humus here is extraordinary—millennia of decay. Imagine the rain of organic matter; needles, spores, twigs, and bugs building the soil from the top down. Photosynthesis—now isn't that a pure wonder, how light and air can end up as soil?" He liked to remind us that the genus *Homo* is derived from the Latin word for "humus" and for "humility." He was a humble man, a man of the soil who signed his name not with PhD appended, but as Francis D. Hole, TNS. It always drew the question.

Dr. Hole passed away many years ago, but a field trip to tree heaven would be heaven for him as well. Field trips with him were legendary, as he introduced us to the art and science of soil horizons, to hardpans, gley layers, and podzols. Known as the "poet laureate of soil," he was fond of quoting Whitman's line "the press of my foot to the earth springs a hundred affections." His affection left behind generations of soil scientists and a folio of soil songs, including ditties like "Oh give me a home, on a deep mellow loam."

A gifted teacher, he played his violin in class so that Bach could tell us stories of soil that words could not.

We stroll among the logs, arm in arm. I stop to stroke the silky mosses on a log, and the palm of my hand is arrested by a sharp metal edge, just beneath the carpet, where an aluminum tag is anchored to the log. A ten-digit number is pressed into the metal. Further down the path, I see orange survey flags rising out the duff. Here, the logs support more than a blanket of moss. Clear plastic tubing emerges from the end of one log. Cylinders of bright PVC pipe are driven into others. Clearly, we are not the first to stumble into forest heaven. The path is well worn by other scientists, who trek to this site to participate in an experiment designed to last two hundred years.

At the time that this experiment was established, the cutting of ancient forests was at its peak. Loaded log trucks careened down the highway in this valley in the Cascade Mountains of Oregon at a rate of one per minute. At the same time that these forests were disappearing, our knowledge of their importance was growing. When a forest was valued only as a commodity, old forests were deemed unproductive, fallen timber, a liability. But forest ecologists at the Andrews Experimental Forest envisioned something vital in ancient forests, essential to the biogeochemical cycles of the earth. So, despite resistance, they initiated long-term ecological research to understand the contributions of old growth and decomposition to the flourishing of air, soil, water, and life—before it was too late. Studying the long process of decay of old growth logs is a doorway to understanding the ties between the elements of earth, air, fire, and water that converge to clothe these mountains in ancient forest.

Dr. Hole breathed in the fecund fragrance of fungi as he swept his arm all around the whole bouquet of color and reminded me that the mushroom garden is only a tiny fraction of the fungal whole. The ground itself, acres wide and feet deep, is woven together with hyphae, the threads of the fungal fabric.

"Come here," he said. We knelt in the duff, and he gently brushed away the needles, exposing the black soil. "Look, you can easily see the mycelium." We saw fans of white threads, rubbery black cords, and tangled webs of yellow lace, all binding together the humus.

A log is a great warehouse of carbon stored away for the long haul. He breaks off a piece of old wood and hands me a magnifying glass. The wood is light in my hand and as full of tiny pores as a sponge. Wood is a mass of hollow vessels, empty spaces that used to conduct water but have become a maze of tunnels for fungal hyphae surrounded on all sides by potential food. But cellulose is far too large a molecule to pass through the cell membrane of the fungus. It simply won't fit. Fungi can't eat trees any more than we can swallow an entire watermelon in one bite. We must cut it into manageable slices. Fungi process their food by secreting digestive enzymes upon it. The molecular equivalent of forks and knives, the enzymes cut the cell wall, cleaving the chemical bonds to reduce the rigid architecture to a syrup of its constituents. Lacking orifices of any kind, fungi absorb their food through the entire surface of the mycelium. It would be as if we ate by lying naked in a bowl of pasta to absorb it through our skin.

I imagine the hum of bacteria in the rotting wood. They roll like seals, plump and sleek with the sweetened moisture of dissolving cells.

All of this disassembly of wood burns a great deal of energy. Log carbon is transformed to mycelial carbon and in the breath of the fungus is exhaled as carbon dioxide. So, too, the beetles and mites and wavy-legged centipedes are exuding carbon dioxide with every morsel of decay.

That puff of liberated carbon dioxide, newly released on the breeze, resided in the tree for five hundred years before it fell. But where did it alight before that? What was its previous life story? And the one before that? Did it live in the body of a Pacific Giant Salamander? A Trillium blossom? Or was it released on the song of a Nez Perce woman picking red huckleberries on this very hilltop?

I like to imagine the moment of liberation for a molecule of carbon dioxide. Imagine being held tight for centuries in the embrace of an ancient tree, locked up in lignin until... the gasp of a fungus-eating beetle sets you free to become a free-floating molecule, a thing of the air, a part of something vast and fluid. Is that how the spirit leaves the body? Released from the weight of wood, into the afterlife of cedars, there is no boundary between the sacred and the mundane.

Dr. Hole sits down on a log and appreciatively examines the carbon dioxide sampling ports inserted in it. He asks me, "When does a tree become a log?" I'm anticipating a riddle of some kind but the answer seems straightforward. A tree becomes a log when it falls to the ground. "Fine," he says, nodding, "so when does a log become a tree?"

I can see the answer right on the log that is dampening our bums as we sit, a line of tiny hemlock seedlings fully rooted in the softened log, forming what is known as a "nurse log." The decaying log provides ideal germination conditions for small hemlock seeds. Atop the log, they are less likely to be grazed and are already closer to the sunlight. The mossy blanket keeps their roots moist during the summer drought. In old forests, massive hemlocks often stand in straight lines—revealing their shared origins on a now long-decayed nurse log. Trees become logs and logs become trees. In the afterlife of cedars, death is just transition, a rearrangement of carbon from one species to the next.

Dr. Hole sometimes referred to trees as "extensions of soil." They stand as mediators or bridges along the continuum of mineral, humus, water, air, and living beings that make a living soil. Every being is an extension of soil, I suppose. What a wondrous thing. We humans emerge, walk around as if we were one thing, willingly oblivious to our true nature, and then we dissolve in order to emerge again in a wholly different form. Dissolution and reunification, forever and ever, amen.

It is, as I understand it, a tenet of most religions—including the ones adhered to at the little gospel church on the highway—that there is some kind of life after death. To this premise, the logs say yes. Logs, dead trees, are more animate when dead than when alive.

We scientists insist that the province of science is not to study the existence of the afterlife. The research plan is designed to measure nutrient flux and changing densities of microbial populations. They will chart the transitions of cellulose to air, of lignin to humus. The goal is to track the fate of all the logs' carbon as it disseminates into the broader ecosystem, to follow the nitrogen from its source in the log to its incorporation to soil, to beetles, to thrushes. The forty-nine-page decomposition study plan is to evaluate the "internal transformation of coarse woody debris" of the log, through its stages of leaching, fragmentation, transport, collapse, and settling. The data sheets are already prepared for the scientists who will follow and complete these measurements long after the designers of the experiment are humus themselves. This is clearly ecological science and not theology. And yet, the end of this experiment will be, I think, experimental theology. It seems to me an act of faith to set up an experiment that will take two hundred years.

And yet—have they not discovered the nature of the afterlife? Perhaps ecology is also experimental theology. Call it internal transformation, call it rot, or call it transcendence. In the afterlife of cedars, nothing is ever dead.

In his long-ago classes, Dr. Hole's teaching stretched our rudimentary understanding of soil, from an inert growth medium to a vibrant convergence of the primal elements: earth, air, fire, and water. The living soil is both the cradle and the grave.

I look up from my examination of a fir seedling and see that Dr. Hole is fading away, or more accurately circling around—his carbon, his minerals, on their way to be with someone else, to live another life. I wave goodbye but I know he'll be back. I catch a line of one of his songs as he goes:

In myself are entwined
flesh and spirit, well inclined.

Dust I am with gift of breath
I feel safe with life and death.

He always signed his name Dr. Francis D. Hole, TNS, waiting
for the question and giving his answer with a twinkle in his eye.
Francis D. Hole, Temporarily Not Soil.

The fluffy comforter of moss invites me to lie down, to nestle
in the curve of a buttressed root, of a tree not yet a log. The humus
conforms to my shape. I could easily stay here. I want to. The moss
feathers around my head and brushes my cheek. I turn over with
nose to tiny fronds and breathe warm and wet into the moss. So
close, I can see my moisture condense on the cool leaf in luminous
drops. Chlorophyll beckons with lovers' embrace and my carbon
dioxide falls into its arms, woman becoming moss. At the very same
moment, the leaf's green sigh of oxygen goes straight to my blood.
Red answers green in the dance of chlorophyll and hemoglobin.
Plant breath becomes animal breath, animal becomes plant, plant
becomes fungus, fungus becomes plant, mouth-to-mouth resusci-
tation that fulfills our deepest longing for union with the earth.

Permissions

Acknowledgments

Our gratitude runs deep for the community of kin who made this series possible. Strachan Donnelley, the founder of the Center for Humans and Nature, was animated and inspired by big questions. He liked to ask them, he enjoyed following the intellectual and actual trails where they might lead, and he knew that was best done in the company of others. Because of this, and because Strachan never tired of discussing the ancient Greek philosopher Heraclitus, who was partial to Fire, we think he would be pleased by the collective journey represented in *Elementals*. One of Strachan's favorite terms was "nature alive," an expression he borrowed from the philosopher Alfred North Whitehead. The words suggest activity, vivaciousness, generous abundance—a world alive with elemental energy: Earth, Air, Water, Fire. We are a part of that energy, are here on this planet because of it, and the offering of words given by our creative, empathic, and insightful contributors is one way that we collectively seek to honor *nature alive*.

A well-crafted, artfully designed book can contribute to the vitality of life. For the mind-bending beauty of the cover design, cheers to Mere Montgomery of LimeRed; she is a delight to work with and LimeRed an incredible partner in bringing to visual life the Center for Humans and Nature's values. For an eye of which an eagle would be envious, a thousand blessings to the deft manuscript editor Katherine Faydash. For the overall style and subtle touches to be experienced in the page layout and design, we profoundly thank Riley Brady. We also wish to thank Ronald Mocerino at the Graphic Arts Studio Inc. for his good-natured spirit and attention to our printing needs, and Chelsea Green Publishing for being excellent collaborators in distribution and promotion.

Thank you to our colleagues at the Center for Humans and Nature, who are elemental forces in their own rights, including our president Brooke Parry Hecht, as well as Lorna Bates, Anja Claus, Katherine Kassouf Cummings, Curt Meine, Abena Motaboli, Kim Lero, Sandi Quinn, and Erin Williams. Finally, this work could not move forward without the visionary care and support of the Center for Humans and Nature board, a group that carries on Strachan's legacy in seeking to understand more deeply our relationships with *nature alive*: Gerald Adelmann, Julia Antonatos, Jake Berlin, Ceara Donnelley, Tagen Donnelley, Kim Elliman, Charles Lane, Thomas Lovejoy, Ed Miller, George Ranney, Bryan Rowley, Lois Vitt Sale, Brooke Williams, and Orrin Williams.

—**Gavin Van Horn and Bruce Jennings**
series coeditors

My gratitude for Shkakami-Kwe (the Earth) and all that she gives, eternal and without condition. I am thankful to learn from each of the contributors through their stories and voices. Many thanks to the series editors for all of their collaborative skills and gratitude for the impact of CHN, an organization that puts great people and opportunities together.

— **Kristi Leora Gansworth, editor**

The Earth collection wouldn't exist without the brilliant editors and contributors who shared their visions and words here, and I extend gratitude to each of them. I would also like to thank my family: Ted, who makes everything possible, and O, who makes everything sweet. I am grateful to live among the memories and dreams of my parents and grandparents, and to be surrounded by the places, plants, animals, and waters of New York City, the center of the Lenape diaspora and the nexus of many Indigenous and immigrant communities from around the world.

— **Hannah Eisler Burnett, editor**

Contributors · volume i

Marcia Bjornerud (she/her) is Professor of Geosciences at Lawrence University in Wisconsin. Her research focuses on the physics of earthquakes and mountain building, and she combines field-based studies of bedrock geology with quantitative models of rock mechanics. She has done research in High Arctic Norway (Svalbard) and Canada (Ellesmere Island), as well as mainland Norway, Italy, New Zealand, and the Lake Superior region. She is the author of several books for popular audiences: *Reading the Rocks: The Autobiography of the Earth*, *Timefulness: How Thinking Like a Geologist Can Help Save the World*, and *Geopedia: A Brief Compendium of Geological Curiosities*.

Hannah Eisler Burnett (she/they) is an anthropologist whose work focuses on water, property, toxicity, and capital. She is the Jamaica Bay Coastal Resilience Specialist for New York Sea Grant, where she works with New York City's coastal communities to imagine and build a just climate future. Before that, she was Communications and Editorial Associate for the Center for Humans and Nature. Hannah received her PhD in anthropology from the University of Chicago in 2023.

Imani Jacqueline Brown (she/her) is an artist, activist, writer, and researcher from New Orleans, based between New Orleans and London. Her work investigates the "continuum of extractivism," which spans from settler-co-lonial genocide and slavery to fossil-fuel production and climate change. In exposing the layers of violence and resistance that form the foundations of settler-colonial society, she opens space to imagine new paths to ecological reparations. Imani combines archival research, ecological philosophy, cultural and legal theory, people's history, and countercartographic strategies to unravel the spatial logics that make geographies, unmake communities, and break Earth's geology.

Nickole Brown (she/her) is the author of *Sister* and *Fanny Says*. She lives in Asheville, North Carolina, where she volunteers at several animal sanctuaries. *To Those Who Were Our First Gods*, a chapbook of poems about these animals, won the 2018 Rattle Prize, and her essay-in-poems, *The Donkey Elegies*, was published by Sibling Rivalry Press in 2020. She's the President of the Hellbender Gathering of Poets, an annual environmental literary festival in Black Mountain, North Carolina.

Photo credit:
Tosca Ophelia

Franny Choi (she/they) is the author of three poetry collections: *The World Keeps Ending, and the World Goes On* (Ecco, 2022), *Soft Science* (Alice James Books, 2019) and *Floating, Brilliant, Gone* (Write Bloody Publishing, 2014). Her writing has appeared in the *New York Times*, *The Atlantic*, the *Paris Review*, and elsewhere. A recipient of the Lily/Rosenberg Fellowship, Princeton's Holmes National

Poetry Prize, and the Elgin Award, Franny is faculty in literature at Bennington College and the founder of Brew & Forge.

Rita Dove (she/her) won the 1987 Pulitzer Prize for her third book of poetry, *Thomas and Beulah*, and served as US Poet Laureate from 1993 to 1995. She is the only poet to receive both the National Humanities Medal and the National Medal of Arts. Recent honors include the Wallace Stevens Award, the American Academy of Arts & Letters' Gold Medal in poetry, the Ruth Lilly Poetry Prize, and the Bobbitt Prize for lifetime achievement from the Library of Congress. Dove teaches creative writing at the University of Virginia; her latest poetry collection, *Playlist for the Apocalypse*, appeared in 2021.

Kristi Leora Gansworth (she/her) is Anishinabe-kwe, a scholar, maker, and poet.

Emma Gilheany (she/her) is an environmental anthropologist and archaeologist of the contemporary and recent past. She is a PhD candidate in anthropology at the University of Chicago, and her dissertation takes place in Nunatsiavut, tracing Inuit sovereignty as a practice. She uses archaeological, ethnographic, archival, and multimedia methodologies to rethink how resistance to imperialism has been theorized using the ephemeral material record of the circumpolar north. She is particularly interested in using archaeological epistemologies to intersect with and serve

Nunatsiavummiut sovereignty and is committed to public and community-engaged scholarship.

Alexis Pauline Gumbs (she/they) is a cherished black feminist vessel of love and aspires to be your favorite cousin. She is the author of several books, most recently *Undrowned: Black Feminist Lessons from Marine Mammals*. Alexis is the recipient of numerous honors including recently, the Windham Campbell award in poetry, the Whiting Award in nonfiction and a National Endowment of the Arts award for her work on her forthcoming biography *Survival Is a Promise: The Eternal Life of Audre Lorde*.

Liam Heneghan is a Dublin-born writer living in the US Midwest.

Jessica Jacobs (she/her) is the author of *unalone*, poems in conversation with the book of Genesis (Four Way Books, 2024), as well as *Take Me with You, Wherever You're Going* (Four Way Books), winner of the Devil's Kitchen and Goldie Awards; *Pelvis with Distance* (White Pine Press), winner of the New Mexico Book Award and a finalist for the Lambda Literary Award; and *Write It! 100 Poetry Prompts to Inspire* (Spruce Books/PenguinRandomHouse), coauthored with Nickole Brown. She is the founder and executive director of Yetzirah: A Hearth for Jewish Poetry.

Danielle B. Joyner (she/her) is Associate Professor of Medieval Art History at Lawrence University in Appleton, Wisconsin. Her current work explores connections between arts and the environment in the European Middle Ages.

Robin Wall Kimmerer, PhD, is a mother, scientist, writer, as well as Distinguished Professor at SUNY College of Environmental Science and Forestry in Syracuse, New York, and the founding director of the Center for Native Peoples and the Environment. She is an enrolled member of the Citizen Potawatomi Nation and a student of the plant nations. Her writings include *Gathering Moss*, which was awarded the John Burroughs Medal for Nature Writing, and the bestselling *Braiding Sweetgrass: Indigenous Wisdom, Scientific Knowledge and the Teachings of Plants*. In 2022, she was awarded a MacArthur Fellowship and was elected to the National Academy of Sciences. As a writer and a scientist, her interests include not only the restoration of ecological communities but also the restoration of our relationships to land.

Laticia McNaughton is enrolled in the Six Nations of the Grand River Mohawk Nation and is a member of the Wolf Clan. She was raised in the Tuscarora Nation community in Western New York. She loves nature and is a fluent Mohawk language speaker, a creative, gardener, cook, and photographer. Laticia is also a PhD Candidate in American Studies at the University at Buffalo. Her dissertation research examines Haudenosaunee (Iroquois) food history, food sovereignty practices, and wellness traditions. She resides in Buffalo, New York, with her husband and two cats.

Raised among the Mashpee Wampanoag Tribe, **Tia Pocknett** (she/her) is a Mi'kmaq/ L'nu woman. She is a mother, wife, teacher, language keeper, and works with the Mashpee Wampanoag Tribe at their tribal school, Weetumuw Katnuhtôhtâeekamuq (Weetumuw School). Tia is currently the lead lower elementary teacher and works with her team to continue to develop a curriculum around culture and language for the school. When she isn't working at Weetumuw Katnuhtôhtâeekamuq, Tia enjoys beading, cooking, foraging, and spending time with her family.

Oyah Beverly A. Reed Scott (she/her/they) is a writer, poet, healer, and gardener who has triumphed over adversity through the power of her open-hearted spirit. As a writer, she shares her personal journey of resilience and growth through story and poetry. Her "Soul Garden" is a literal symbol of transformation, source of spiritual and ancestral communion, and safe space for healing for those who resonate with her calling. She believes "love will win" and honors the beauty and resilience of the human spirit. Her work reminds us to, no matter the challenges we face, keep going.

Jane Slade (she/her), MID, LC, IES, is a lighting researcher, writer, and speaker at Anatomy of Night on the impacts of light upon the environment, wildlife, and human health. She is also the host of the podcast *Starving for Darkness.* Jane was awarded the International Dark-Sky Association's 2021 Dark Sky Defender for North America and was a Richard Kelly Grant recipient for explorations into the social and emotional impacts of light and lighting. She is a member of the IES Outdoor Nighttime Environment (ONE) Committee and the

IES Progress Report Committee, is a contributor to LD+A on the topic of wildlife, and is currently writing a book about the natural daylight cycle.

Melissa Tuckey (she/her) is author of *Tenuous Chapel,* a book of poems selected by Charles Simic for ABZ Press's First Book Award and *Ghost Fishing: An Eco-Justice Poetry Anthology.* She's a former Tompkins County Poet Laureate and an emeritus fellow at Black Earth Institute. Other honors include a Fine Arts Work Center Winter Fellowship and grants from Ohio Arts Council and DC Commission on the Arts and Humanities. Melissa currently lives in Ithaca, New York, land that is the traditional homeland of Gayogohó:nǫʔ people, where she writes, gardens, and teaches.

Andreas Weber (he/him), PhD, is a biologist, philosopher, and nature writer. He has published more than a dozen books. His most recent English-language books are *Enlivenment: Toward a Poetics for the Anthropocene* (2019) and *Sharing Life: The Ecopolitics of Reciprocity* (2020). Andreas lives in Berlin and in Italy's Apennine mountains.